symposia on theoretical physics

4

Contributors to this volume:

L. A. P. Balazs
R. Blankenbecler
V. Devanathan
K. Dietz
Harish - Chandra
P. T. Landsberg
J. Lukierski
Ph. Meyer
A. N. Mitra
D. J. Morgan
V. Singh
A. Sjolander
S. K. Srinivasan
M. H. Stone
K. Venkatesan
V. Weisskopf

symposia on theoretical physics

Lectures presented at the
1965 Third Anniversary Symposium
of the Institute
of Mathematical Sciences
Madras, India

Edited by
ALLADI RAMAKRISHNAN
Director of the Institute

PLENUM PRESS • NEW YORK • 1967

ISBN-13: 978-1-4684-7693-4   e-ISBN-13: 978-1-4684-7691-0
DOI: 10.1007/ 978-1-4684-7691-0

Library of Congress Catalog Card Number 65-21184

Introduction

The Third Anniversary Symposium, held in January 1965, was devoted mainly to various topics in elementary particle physics, with a few lectures on many-body problems and a short supplementary program in mathematics.*

In the Introductory Address Professor V. Weisskopf, Director-General of CERN, Geneva, presented a broad survey of the then current scene in elementary particle physics, the most dominant trend in which is the concept of symmetry. He traced the use of the concept of rotational invariance and symmetry under permutation of identical objects in the realm of atomic spectra and how, with the inclusion of isotopic spin, such use was extended to the study of properties of nuclei. Professor Weisskopf also described how, in addition, elementary particles are characterized by a new quantum number, the hypercharge, which, with isotopic spin, is part of a wider symmetry $SU(3)$. He mentioned three classes of experiments at CERN, one in search of quarks, one to investigate the existence of vector bosons suggested by theories as possible mediators of weak interaction, and one to test the existence of cosmic forces to explain C P or T violation. The quotations from Newton's *Opticks* at the beginning and the end of the lecture were strikingly relevant.

Two lectures dealt with the application of $SU(3)$ symmetry to weak and strong interactions, respectively. Ph. Meyer of the University of Paris, Orsay summarized his work on the conserved vector current hypothesis in relation to broken symmetries. By assuming that the violation of $SU(3)$ invariance can be described by a local

---

*While this introduction discusses all the lectures delivered at the Symposium at Matscience in January 1965, it has not been possible to include all of them in this volume.

Lagrangian which transforms like a member of an $SU(3)$ multiplet with isospin and strangeness equal to zero, he proved that the first order correction to the vector decay amplitude in the limit of zero momentum transfer can be accounted for by using unrenormalized coupling constants but with wave functions corresponding to the physical masses. The multiplet assignments of various observed particles in the $SU(6)$ scheme (in which one fuses the internal symmetry group for the hadrons with the ordinary spin) and the different sum rules between the masses of the hadrons which are in good agreement with the observed masses, were described by V. Singh of the Tata Institute of Fundamental Research, Bombay.

The concept of an equivalent potential in quantum field theory and S-matrix theory and the use of a nonlocal potential in calculations in elementary particle physics formed the subject matter for three talks. R. Blankenbecler of Princeton University showed how one could obtain upper and lower bounds for the phase shifts and the K-matrix elements in nonrelativistic problems. He then extended these ideas to the relativistic case where, starting from the Bethe–Salpeter equation and including multiparticle states, a potential can be constructed in a nonperturbative fashion. Application of the method to the ρ-meson bootstrap problem does not lead to any self-consistent solution. L.A.P. Balazs of the Tata Institute discussed a generalization of the work of Charap and Fubini in which the potential is constructed by requiring that it reproduce the relativistic amplitude at any energy, and is obtained by calculating the absorptive parts in the crossed channel reactions for increasingly larger values of the momentum transfer by interactions using the strip approximation to the Mandelstam representation. Starting from a nonlocal potential corresponding to a repulsive interaction, A. N. Mitra of the University of Delhi explained how by using the potential in a Schrödinger-type equation a detailed examination of the phase shifts can lead to an understanding of some pion resonances. A model for peripheral interactions below 10 GeV in which the K-matrix element for quasi two-particle reactions were replaced by the corresponding Born terms while the remaining K-matrix elements for higher particle final states were assumed to have a statistical distribution with zero mean value was presented by K. Dietz of CERN, Geneva.

J. Lukierski of the University of Wroclaw, Poland, presented some considerations regarding the renormalizability of theories of particles of spin higher than, or equal to, ones which are traditionally considered to be unrenormalizable. He showed how the choice of suitable projection operators may lead to renormalizability conditions less stringent than the usual ones.

A possibility of renormalizing theories (considered usually unrenormalizable) using Caianiello's approach involving Pfaffians and Hafnians was suggested by N. R. Ranganathan and R. Vasudevan of Matscience, Madras. M. Gourdin of Orsay, France, gave a compact formula for calculating the matrix element of electron scattering from a target of arbitrary spin in a covariant way, and K. Venkatesan of Matscience, Madras, explained how the notion of a group representation breaks down for certain values of the group parameter in the case of complex angular momentum.

The three lectures on many-body problems started with that of C. de Dominicis of Saclay (France) who dealt with the quasiparticle formulation of quantum statistical mechanics which is based on partial summations in diagrammatic expansions and discussed the relationship with the Landau theory of Fermi liquids. A method of obtaining sum rules for any system, given the Hamiltonian of the system and the main variables which one wants to occur in the sum rules, described by P. T. Landsberg and D. J. Morgan of the University College of Cardiff, Wales. A. Sjölander of the Institut för Mekanik in Göteborg, Sweden, gave an account of the concept of lattice waves or phonons and described in detail the inelastic neutron scattering technique and its theory as a means by which the experimental determination of phonon dispersion curves and polarization directions are exclusively done at present.

In the supplementary mathematics session, M. H. Stone of the University of Chicago, emphasized the need for enquiry into the techniques of model construction in the various mathematical sciences and the study of the role of the mathematician in this respect. We are particularly grateful to Professor Harish-Chandra, at the Institute for Advanced Study in Princeton, for his active participation in the symposium by giving a series of two lectures on the problem of constructing the characters for noncompact, semi-simple Lie groups. S. K. Srinivasan of the Indian Institute of Technology, Madras, gave a brief outline of some recent developments in stochastic point processes.

It is our earnest hope that the supplementary sessions in mathematics will soon grow into a full-fledged symposium to be conducted concurrently with that on theoretical Physics.

Alladi Ramakrishnan

Contents

Contents of Other Volumes xi

Introductory Address... 1
>by V. Weisskopf, *CERN, Geneva, Switzerland*

Conserved Vector Currents and Broken Symmetries............ 13
>by Ph. Meyer, *University of Paris, Orsay, France*

Multiplet Structure and Mass Sum Rules in the *SU*(6)
Symmetry Scheme 23
>by V. Singh, *Tata Institute of Fundamental Research, Bombay,
>India*

A New Approach to Scattering Theory 35
>by R. Blankenbecler, *Princeton University, Princeton, New
>Jersey*

Equivalent Potential Approach for Strong Interactions 43
>by L. A. P. Balazs, *Tata Institute of Fundamental Research,
>Bombay, India*

"Repulsive" Potential Approach to Pion Resonances........... 51
>by A. N. Mitra, *University of Delhi, Delhi, India*

A Model of a Unitary *S*-Matrix for Peripheral Interactions..... 59
>by K. Dietz, *CERN, Geneva, Switzerland*

The Renormalizability of Higher Spin Theories............... 69
>by J. Lukierski, *University of Wroclaw, Wroclaw, Poland*

Group Representations for Complex Angular Momentum....... 91
>by K. Venkatesan, *Matscience, Madras, India*

Muon Capture by Complex Nuclei......................... 103
>by V. Devanathan, *University of Madras, Madras, India*

Comments on Sum Rules 113
>by P. T. Landsberg and D. J. Morgan, *University, College,
>Cardiff, Wales*

Inelastic Neutron Scattering and Dynamics in Solids and
Liquids 121
　　by A. Sjölander, *Institut för Mekanik, Göteborg, Sweden*
Axioms and Models 131
　　by M. H. Stone, *University of Chicago, Chicago, Illinois*
Characters of Semi-Simple Lie Groups 137
　　by Harish-Chandra, *Institute for Advanced Study, Princeton,
　　New Jersey*
Sequent Correlations in Evolutionary Stochastic Point Processes. 143
　　by S. K. Srinivasan, *Indian Institute of Technology, Madras,
　　India*
Author Index 157
Subject Index 159

Contents of Other Volumes

VOLUME 1

Symmetries and Resonances
  T. K. Radha

Group Symmetries with *R*-Invariance
  R. E. Marshak

Regge Poles and Resonances
  T. K. Radha

On Regge Poles in Perturbation Theory
  and Weak Interactions
  K. Raman

Determination of Spin-Parity
  of Resonances
  G. Ramachandran

Pion Resonances
  T. S. Santhanam

Pion–Nucleon Resonances
  K. Venkatesan

The Influence of Pion–Nucleon
  Resonance on Elastic Scattering
  of Charged Pions by Deuterons
  V. Devanathan

Pion–Hyperon Resonances
  R. K. Umerjee

Some Remarks on Recent Experimental
  Data and Techniques
  E. Segre

On New Resonances
  B. Maglić

The Higher Resonances in the
  Pion–Nucleon System
  G. Takeda

VOLUME 2

Origin of Internal Symmetries
  E. C. G. Sudarshan

Construction of the Invariants of the
  Simple Lie Groups
  L. O'Raifeartaigh

On Peratization Methods
  N. R. Ranganathan

Large-Angle Elastic Scattering at
  High Energies
  R. Hagedorn

Crossing Relations and Spin States
  M. Jacob

The Multiperipheral Model for
  High-Energy Processes
  K. Venkatesan

Regge Poles in Weak Interactions and
  in Form Factors
  K. Raman

Some Applications of Separable
  Potentials in Elementary Particle
  Physics
  A. N. Mitra

Form Factors of the Three-Nucleon
  Systems H^3 and He^3
  T. K. Radha

Muon Capture by Complex Nuclei
  V. Devanathan

Electrodynamics of Superconductors
  B. Zumino

"Temperature Cutoff" in Quantum
  Field Theory and Mass
  Renormalization
  S. P. Misra

Recent Developments in the Statistical
  Mechanics of Plasmas
  H. DeWitt

Effective-Range Approximation Based
on Regge Poles
B. M. Udgaonkar

Some Current Trends in Mathematical
Research
M. H. Stone

Semigroup Methods in Mathematical
Physics
A. T. Bharucha-Reid

Introduction to Quantum Statistics of
Degenerate Bose Systems
F. Mohling

Recent Mathematical Developments in
Cascade Theory
S. K. Srinivasan

Theory of a General Quantum System
Interacting with a Linear Dissipation
System
R. Vasudevan

VOLUME 3

Many-Particle Structure of Green's
Functions
K. Symanzik

Broken Symmetries and Leptonic
Weak Interactions
R. Oehme

Partial Muon Capture in Light Nuclei
A. Fujii

Quantum Gauge Transformations
J. Lukierski

Theories of Particles of Arbitrary Spins
K. Venkatesan

Bethe–Salpeter Equation and
Conservation Laws in Nuclear
Physics
W. Brenig

On a Class of Non-Markovian
Processes and Its Application
to the Theory of
Shot Noise and Barkhausen Noise
S. K. Srinivasan

VOLUME 5

Lectures on Nested Hilbert Spaces
A. Grossmann

Elementary Homology Theory
N. R. Ranganathan

Application of Algebraic Topology
to Feynman Integrals
V. L. Teplitz

Weak Interactions
R. J. Oakes

Fundamental Multiplets
A. Ramakrishnan

The Ground States of He^3 and H^3
K. Ananthanarayanan

Remarks of the Present State of
General Relativity
S. Kichenassamy

On the Conformal Group and its
Equivalence with the Hexa-
dimensional Pseudo Euclidean
Group
M. Baktavatsalou

Certain Extremal Problems
K. R. Unni

Fluctuating Density Fields and
Fokker–Planck Equations
S. K. Srinivasan

Contraction of Lie Groups and
Lie Algebras
K. Venkatesan

A Pattern in Functional Analysis
J. L. Kelley

Correspondence Principles in
Dynamics
R. Arens

VOLUME 6

On Locally Isomorphic Groups and Cartan–Stiefel Diagrams
B. Gruber

Linear Response, Bethe–Salpeter Equation, and Transport Coefficients
L. Picman

The Description of Particles of Arbitrary Spin
P. M. Mathews

Radiative Corrections in β-Decay
G. Källén

What Are Elementary Particles Made Of?
E. C. G. Sudarshan

Recent Developments in Cosmology
J. V. Narlikar

An Introduction to Nevanlinna Theory
W. K. Hayman

Normalization of Bethe–Salpeter Wave Functions
Y. Takahashi

Non-Lagrange Theories and Generalized Conservation
Y. Takahashi

Cosmic X-Rays, γ-Rays, and Electrons
R. R. Daniel

β-Decay and μ-Capture Coupling Constants
S. C. K. Nair

Functions of Exponential Type
K. R. Unni

A Model for Processing Visual Data with Applications to Bubble Chamber Picture Recognition (Summary)
R. Narasimhan

On Functional Methods in the S-Matrix Theory
J. Rzewuski

An Impact Parameter Formalism
T. Kotani

Some Properties of the Fourier–Bessel Transform
G. Källén

The A_1 and K^{**} (1320) Phenomena—Kinematic Enhancements of Mesons?
G. and S. Goldhaber

The Photoproduction and Scattering of Pions from H^3 and He^3
K. Ananthanarayanan

Relativistic Extensions of $SU(6)$
H. Ruegg

A Survey of π-N Scattering, and of the $T = \frac{1}{2}$ Amplitudes
B. J. Moyer

Introductory Address

VICTOR WEISSKOPF

DIRECTOR-GENERAL, CERN
Geneva, Switzerland

The science of elementary particles has in recent years naturally become the most fundamental of all pursuits in physics. In this lecture, I shall attempt to present a broad and distant view of the current scene in elementary particle physics and indicate to you the general trend of ideas in this field. The most dominant in this trend is the concept of symmetry, which is playing an increasingly fundamental role in the study of interactions of elementary particles. It will be rewarding for us to discuss this in an historical perspective. We shall start with a quotation from Newton's *Opticks*:

> "All these things being consider'd, it seems probable to me, that God in the Beginning form'd Matter in solid, massy, hard, impenetrable, moveable Particles, of such Sizes and Figures, and with such other Properties, and in such Proportion to Space, as most conducted to the End for which he form'd them; and that these primitive Particles being Solids, are incomparably harder than any porous Bodies compounded of them; even so very hard, as never to wear or break in pieces; no ordinary Power being able to divide what God himself made one in the first Creation. While the Particles continue entire, they may compose Bodies of one and the same Nature and Texture in all Ages: But should they wear away, or break in pieces, the Nature of Things, depending on them, would be changed. Water and Earth, composed of old worn Particles and Fragments of Particles, would not be of the same Nature and Texture now, with Water and Earth composed of entire Particles in the Beginning. And therefore, that Nature may be lasting, the Changes of Corporeal Things are to be placed only in the various Separations and new Associations and Motions of these permanent Particles; compound Bodies being apt to break, not in the midst of solid Particles, but where those Particles are laid together, and only touch in a few Points."

We clearly perceive the remarkable vision of Newton in the above quotation. He is obviously worried about elementary particles. He is troubled that elementary particles may lose an edge here or a corner there in course of time. Since nature is of infinite duration, he wonders how the properties of these particles will remain the same. So he wants them to be incomparably hard. Today the elementary particle physicist is able to answer Newton's concern.

Rutherford discovered in the 1920's that atoms are not incomparably hard, and indeed they are composed of electrons and nuclei. It is then hard to understand how the atoms can keep their properties. We are now aware of the phenomenal success of the concepts of quantum mechanics, which provide an explanation of a wide variety of properties of various atoms. If we probe deeper into the cause of this tremendous success, we perceive that the properties of the quantum states are defined by an underlying symmetry. The shape of the wave functions of an electron in a Coulomb field is a good illustration of the situation.

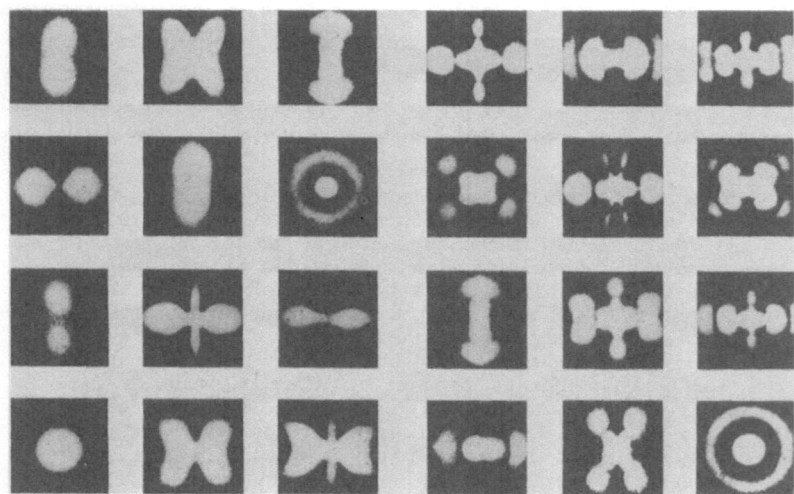

Fig. 1

We now observe that the atomic states possess two distinct symmetries: (a) rotational symmetries, which also dictate spin symmetry and (b) symmetry under permutation because of identity of electrons. The latter is known as the *Pauli principle*. It is this principle which gives the different varieties of atoms distinct properties, since the electrons are forced to occupy various orbital states with different energies.

These symmetries are all related to certain conservation principles. We know that rotational symmetry gives rise to conservation of total angular momentum, while the identity of particles is due to symmetry under permutation. We also believe that the conservation of charge is due to invariance under a gauge transformation. In fact, this invariance is the basis of the light-quantum hypothesis and of our study of quantum electrodynamics. In short, the principles of symmetry described above ensure that the properties of particles are preserved. This remark does in a sense answer Newton's problem.

In our picture of atoms given above, the nuclei seem to be incomparably hard. Thus, Newton's worry merely gets shifted from the atoms to nuclei. It was again Rutherford who soon established that even nuclei are not incomparably hard. They consist of protons and neutrons, and evidently they cannot be elementary particles. How is it that nuclei have characteristic unchanging properties? It is remarkable again that the answer should be based on symmetry.

The analogy between atomic physics and nuclear physics is surprisingly closer than we should perhaps expect. We have in nuclear physics shell model nuclear wave functions, a periodic table for describing the properties of nuclei, and so on. If we look more closely into the underlying symmetries, we discover that the nuclear interactions also possess invariance under rotations and permutations. Besides these, there is a new symmetry known as *isospin symmetry* which treats the proton and neutron as two states of a particle called the *nucleon*. Though this symmetry is exact as far as nuclear forces are considered, it is violated by electromagnetic interactions.

Let us look at some relevant numbers in these two systems. A rough estimate of the size of the atoms is provided by the Bohr radius

$$a = \frac{\hbar^2}{mc^2} = 5 \times 10^{-9} \, \text{cm}$$

The interaction energy is measured in Rydberg units. A Rydberg in atomic physics is given by

$$\text{Ry}_{\text{atom}} = \frac{me^4}{\hbar^2} = 27 \, \text{eV}$$

We also observe that the force between nuclei and electrons is proportional to e^2/r. Turning to nuclear physics, we find that the force between nucleons is given by Yukawa interactions

$$\frac{g^2}{r}e^{-r/r_0}$$

where $g^2/\hbar c \sim 0.08$ and r_0 is the range of nuclear forces. This force differs from the previous one by an exponential factor. If, for a moment, we set this factor to be unity, we really have a very good analogy between atomic and nuclear physics. We can define a nuclear radius a_N and a Rydberg (nuclear) as a measure of size of the nuclei and of nuclear interaction energy, respectively. We have

$$a_N = \frac{\hbar^2}{Mg^2} = 2.5 \times 10^{-13}\,\text{cm}$$

where M is the mass of the nucleon, and

$$\text{Ry}_{\text{nuclear}} = \frac{Mg^4}{\hbar^2} = 6\,\text{MeV}$$

Since the range of nuclear forces r_0 is 2×10^{-13}cm, we find that the analogy between nuclear and atomic physics can be used in a qualitative manner. If Newton were alive today, he would have been pleased about the fundamental role of symmetry in understanding the "God-given" shapes of atoms and nuclei.

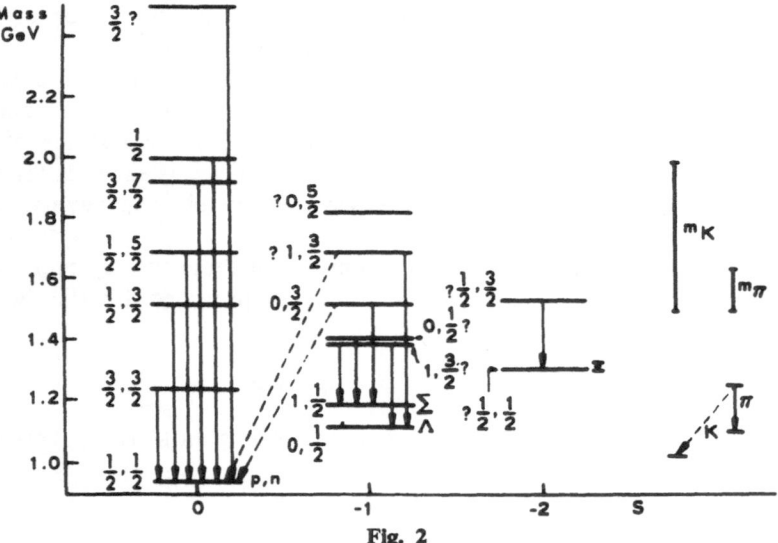

Fig. 2

We now have three particles, the proton (p), neutron (n), and electron (e). Are these incomparably hard? Now we are leaving the domain of reasonably well-formed knowledge, and our only guide is the data emerging from experiments. We are merely tapping in the dark. There exists now strong evidence to show that p and n are not hard enough, and, in fact, we find excited states of p and n. Let us look at the spectrum of baryons in Fig. 2.

For the sake of comparison, let us examine typical atomic and nuclear spectra (Fig. 3).

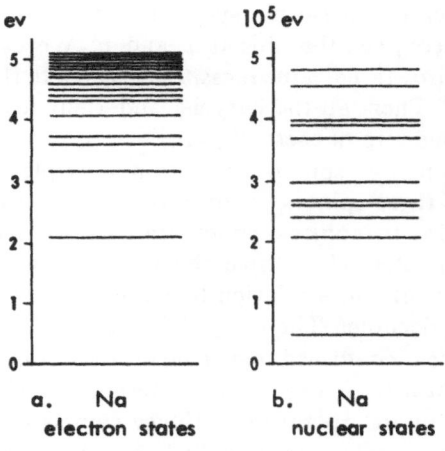

Fig. 3

There are now three types of spectroscopy, each with its characteristic level spacing and characteristic quantum numbers. In the case of elementary particles, we have rotational and isospin symmetries. But there is also a new symmetry, called *hypercharge symmetry*. As we should expect, there are now new selection rules because of new quantum numbers. In atomic spectroscopy, light quanta are being emitted and absorbed. In elementary particle spectroscopy, it is pions (π) and kaons (K) that are emitted and absorbed. Light quanta have only angular momentum, while π and K have charge and mass. Also, K carries a hypercharge. We can consider Λ- and Σ-particles to be metastable states of the nucleon, and perhaps we should not call them elementary particles. Weak interactions do allow Λ and Σ to decay into nucleon states.

Can we succeed again in postulating a new symmetry to understand these spectra? Some regularities in the spectrum of elementary particles soon caught the eye of scientists, and they boldly suggested the SU_3 symmetry, combining the symmetries under isospin and hypercharge. The light baryons were soon grouped into the octet representation of SU_3. More than this, it was suggested that the resonances N^*, Y^*, and Ξ^* form a decuplet. As we are all aware, the missing member Ω in this decuplet was discovered last year at Brookhaven, giving SU_3 symmetry its greatest triumph to date. SU_3 symmetry is only weakly violated in strong interactions.

SU_3 symmetry is merely an extension of isospin symmetry. To see this, we observe that isospin symmetry refers to the two-valuedness of the nucleon, i.e., the objects p and n. We can say that the SU_3 group requires three-valuedness. Let us denote the three objects as p', n', and λ'. Then, all the baryons and their resonances can be thought of as made up of these objects.

If we combine two spinors, we obtain a triplet and a singlet. Combination of three spinors yields a quadruplet. In similar fashion, if we combine two objects which are part of a triad, we obtain a singlet and an octet, while three objects yield an octet and a decuplet. We can work out a relation between the masses of members in an octet or a decuplet. The mass of Ω predicted by such a mass formula has been confirmed experimentally. This indicates that there is some truth in SU_3 symmetry. Recently, SU_3 has been enlarged into SU_6 by incorporating ordinary spin as well.

If we turn to bosons, again we find that they have a spectrum; that is, pions and K-mesons do have higher excited states. Can light quanta have excited states? We know that light quanta scatter each other, through intermediate electron-positron pairs. In fact, positronium can be considered as the excited state of two light quanta.

We know that π- and K-particles interact very strongly with baryons, with a coupling constant

$$G^2 = \frac{M^2}{2m\pi^2}g^2 \sim 14$$

The most impressive feature is that these particles can also be grouped into octets, as in the case of baryons. Can we understand the basic reason for this remarkable regularity in the spectra of elementary particles?

There are at present two broad approaches to answering this question. We believe that both of them may not be successful, and perhaps some new idea will have to be developed. The first approach

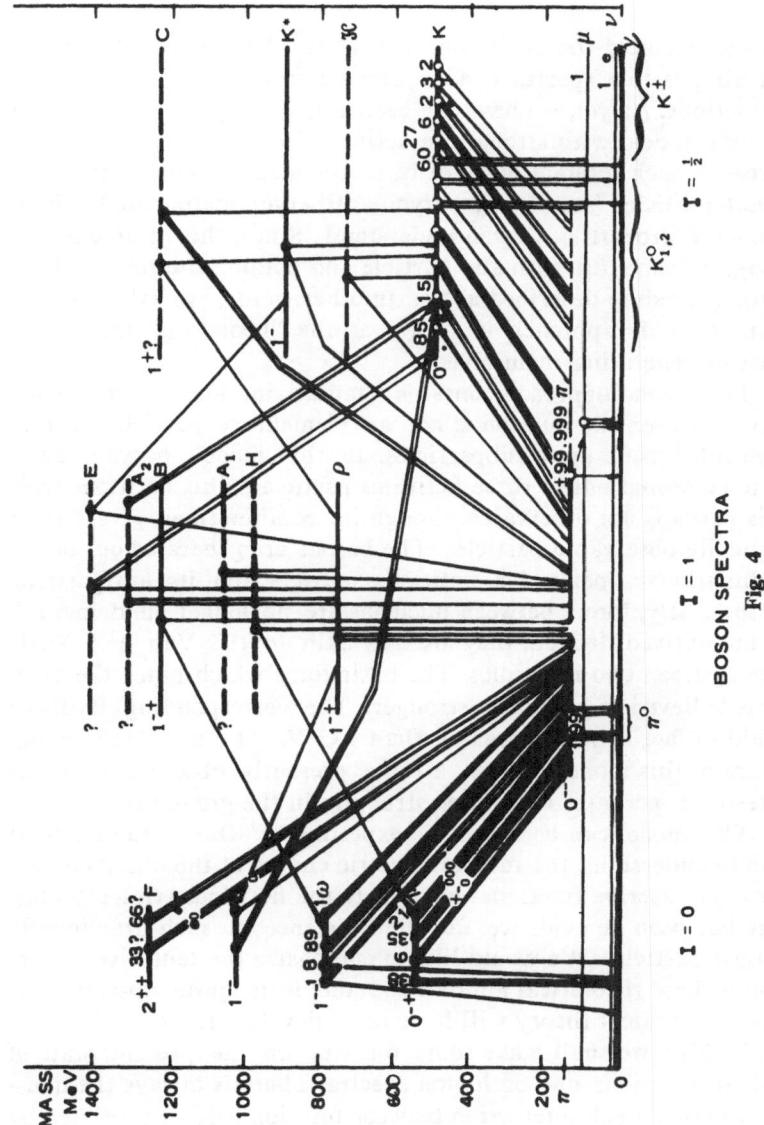

BOSON SPECTRA

Fig. 4

is based on field-theoretic concepts. Here, the inherent belief is that all possible spectra can be derived from a theory of strong interactions. As yet, we have not been able to develop mathematical methods to deal with strong interactions. However, we know that sources of such interactions will be surrounded by some patterns of virtual particles (say, π, K, baryon-antibaryon pairs, etc.) whose symmetry properties we can understand. Since the interaction is strong, we can start with any particle and obtain all other particles by the procedure described above. In other words, everything hangs together. This approach is known as the "bootstrap" theory for particles generating themselves.

The second approach contains a tantalizing idea. Here, we assume that even the nucleon is not an elementary particle. It is a "molecule" made up of subparticles. In this model, baryons arise out of combinations of three fictitious particles. This fictitious triplet is perhaps not observable, though its combinations give rise to physically observable particles. The bosons are generated out of the combinations of pairs of a fictitious particle and its antiparticle. Consequently, forces between nucleons are no longer fundamental, but are derived. Indeed, they are now akin to the Van der Waals force between two molecules. The basic force which binds the triplets is believed to be much stronger. The corresponding Rydberg should in fact be much greater than 1 GeV. The most interesting feature of this model is that all the presently observed particles correspond, perhaps, to the fine structure in the ground state.

This model can be tested by experiments. One of the simplest ways to understand the integral electric charge of the observed particles is to ascribe fractional charges to the fictitious triplet. So far, there has been no evidence for the existence of such fractionally charged particles. We would like to emphasize the tentative nature of both these theoretical approaches, and it is quite possible that some new radical theory will have to be developed.

Finally, we shall make some remarks on the present state of weak interactions and on lepton spectra. There is always the question whether weak interaction between fermion pairs is mediated by a massive and charged vector meson W in a manner analogous to electromagnetic interactions. If this hypothesis is correct, there should be a W bremsstrahlung. So far, experiments have shown that the W-meson is not there if its mass is less than 2 GeV. Precisely speaking, we can only say that either the mass of the W-meson is greater than 2 GeV or it does not exist. These experiments were carried out by means of the strongest available neutrino beam at CERN.

By scattering electrons from protons and deuterons, we are able to determine the electromagnetic form factors of the proton and neutron. We can also study the electron–proton scattering due to weak interactions. Here, we have two form factors, the vector and axial vectors f_V and f_A, respectively. Recent experiments at CERN have revealed that $f_V \sim f_A$ and that they have a shape similar to the electromagnetic form factor. The equality of f_V with the latter was predicted earlier by the conserved vector current hypothesis.

Another interesting result to emerge from recent experiments at CERN concerns the structure of weak interactions. It is assumed that weak interaction is essentially an interaction between currents. Each current contains two terms, one made of the lepton operator (l) and the other from the baryon operator (N). Hence, we have always terms of type $(\bar{N} N)(\bar{l} l)$. We have no precise reason to rule out terms of type $(\bar{N} l)(\bar{l} N)$. Of course, we are aware that a Fierz transformation connects these two terms for vector–axial interaction and, hence, the experimental consequence for small momentum transfer should be the same. However, it should be noted that the momentum transfer in the two cases is very different. The latter form involves high momentum transfer if the two leptons have similar momenta. Recent experiments on weak interaction form factors have shown that they depend only on the momentum transfer in $(\bar{N} N)$ and not on $(\bar{N} l)$. Though this result may not be revolutionary, it is very important, since it confirms our basic hypothesis.

Very recently we have become aware that weak interactions seem to violate not only Y, I, and P but also invariance under charge conjugation and parity transformations. This violation is found to be weak, as is shown by the rate of the 2π-decay mode of K_2^0. Violation of parity conservation alone is not so bad since there is still C P invariance, and we only need to use the antiworld as the mirror world. However, if violation of C P is true, we are in difficulty, because the mirror world does not permit a realizable state in the world.

Perhaps this conclusion is not necessarily warranted by the experiment. Perhaps the interpretation of the experimental result is not correct. It is to be remembered that while the parity violation was maximum, the C P violation is in fact very small. There have been suggestions that this is caused by a new interaction which is much weaker.

Lee, Bernstein, and Bell have brought in a new cosmic interaction to explain the small C P violation. We live in an asymmetric world containing only matter, and perhaps there exists a very weak force which distinguishes between matter and anti matter. It has been shown that if this interaction is taken into account, then $K_2^0 \rightarrow 2\pi$ proceeds and the rate depends on the energy. A crucial experiment to test the hypothesis of cosmic force is now in progress at CERN.

Another development in weak interactions suggested by Cabbibo assumes that the interaction is a vector in a definite direction in SU_3 space. As a consequence, there are definite relations between different weak decays. Many experiments are under way to test such predictions, but it will be some time before the results are available.

At present there are four leptons, electron (e), electron-neutrino (ν_e), muon (μ), and muon-neutrino (ν_μ).

Are we observing the beginning of a fourth spectroscopy? So far, we are facing the intriguing fact that the μ-meson is very much like the electron, except for its mass. We are doing two crucial experiments to test this. The first is a precision experiment on the magnetic moment of μ. The second experiment is designed to observe the lepton pairs $\bar{\mu}\mu$ and $\bar{e}e$ in the annihilation of protons and antiprotons and to test whether there is a difference in probability. Only rarely do lepton pairs occur in the final state, as $\bar{p}p$ always decay into many pions. Since the momentum transfer involved in production of lepton pairs is very high, this is a very difficult experiment.

How can we understand the lepton spectrum? One possibility is to think of a new kind of force. It is quite possible that the electromagnetic force itself may be the cause of the lepton spectrum. At very small distances, the electromagnetic forces are not weak but very strong. These strong forces may very well create a spectrum. These are just speculations.

Does elementary particle physics have any cosmological importance? We sometimes may get the feeling that we are dealing with phenomena which happen only at Berkeley, CERN, and Brookhaven, etc., and perhaps nowhere else in the universe. But if we consider phenomena beyond the earth—in the wider universe—we see that the energetics of interactions are characterized by different levels: the glowing matter from a volcano is caused by atomic processes, and nuclear reactions are responsible for the explosions of supernovae (e.g., the Crab Nebula). With the recent dicovery of quasars, we begin to believe that perhaps sub nuclear processes are taking place in the universe.

There is no better way to summarize this lecture than with another quotation from Newton, which testifies to his remarkable vision:

"Now the smallest Particles of Matter may cohere by the strongest Attractions, and compose bigger Particles of weaker Virtue; and many of these may cohere and compose bigger Particles whose Virtue is still weaker, and so on for divers Successions, until the Progression end in the biggest Particles on which the Operations in Chemistry, and the Colours of natural Bodies depend, and which by cohering compose Bodies of a sensible Magnitude.

"There are therefore Agents in Nature able to make the Particles of Bodies stick together by very strong Attractions. And it is the Business of experimental Philosophy to find them out."

Conserved Vector Currents and Broken Symmetries

PH. MEYER[†]

UNIVERSITY OF PARIS
Orsay, France

I would like to discuss briefly some recent work[1-3] on the problem of the renormalization of the coupling constant of the weak vector currents—essentially the strangeness-changing current—when the symmetry which leads to a conservation law is broken. The main result is that under rather general conditions there is no renormalization to first order in the symmetry-breaking interaction.

Let us first recall well-known facts about the relation between an invariance law, a conserved current, and the nonrenormalization of the corresponding coupling constant. We shall consider a Lagrangian density:

$$L\left(\psi_A, \frac{\partial \psi_A}{\partial x_\mu}\right) \tag{1}$$

which is a function of a certain number of fields ψ_A and their derivatives. By assumption, L is invariant under a set of infinitesimal unitary transformations whose generators F_i have the commutation properties of a Lie algebra:

$$[F_i, F_j] = f_{ijk} F_k \tag{2}$$

Under the group operation F_i, the fields transform as

$$\psi'_A(x, t) = \psi_A(x, t) - i\Lambda_i[F_i, \psi_A(x, t)] \tag{3}$$

$$\psi'_A(x, t) = \psi_A(x, t) + \sum_B M^i_{AB} \psi_B \tag{4}$$

---

†Laboratory of Theoretical and High Energy Physics, Faculty of Sciences.

where Λ_i is an infinitesimal constant associated with the transformation F_i,[†] and the matrices M^i are the basis of a representation of the algebra. The transformed Lagrangian is given by

$$L'(x, t) = L(x, t) + i\Lambda_i \partial_\mu J_\mu^i(x, t) \qquad (5)$$

where

$$J_\mu^i(x, t) = \sum_{A,B} \frac{\partial L}{\partial(\partial \psi_A / \partial x_\mu)} M^i_{AB} \psi_B \qquad (6)$$

Invariance of L under the group implies the conservation law for the currents J_μ^i:

$$\partial_\mu J_\mu^i = 0 \qquad (7)$$

With the usual commutation rules for the fields, we can write for the generators

$$F_i = \int J_4^i d\mathbf{x} \qquad (8)$$

If the currents are conserved, the F_i are independent of time and commute with the total Hamiltonian.

For the isotopic group $SU(2)$, the F_i are the isotopic spin currents; for $SU(3)$, the F_i are the eight unitary spin currents, etc. The electromagnetic current corresponds to the gauge group with $F = Q$, the charge operator.

Now consider the commutation relation

$$[F_i, \psi_B] = -\sum_{B'} M^i_{BB'} \psi_{B'} \qquad (9)$$

Taking the left side between the vacuum and the physical state $|A\rangle$ belonging to the same representation of the group as $|B\rangle$, we get

$$\langle 0|[F_i, \psi_B]|A\rangle = \sum_\alpha \langle 0|F_i|\alpha\rangle\langle\alpha|\psi_B|A\rangle - \sum_\alpha \langle 0|\psi_B|\alpha\rangle\langle\alpha|F_i|A\rangle \qquad (10)$$

Since F_i commutes with H and is a generator of the group, it has nonzero matrix elements only between states of the same representation (with the same momentum). We can rewrite (10) as

$$\langle 0|[F_i, \psi_B]|A\rangle = -\langle 0|\psi_B|B\rangle\langle B|F_i|A\rangle \qquad (11)$$

Comparing with (9), taken between the vacuum and the state $|A\rangle$,

---

[†]We are assuming here that the F_i act on internal degrees of freedom (*I*-spin, *SU*(3)-spin, etc. (and not on the space coordinates).

we obtain

$$\langle B(p_2)|F_i|A(p_1)\rangle = M_{BA}^i \delta(\mathbf{p}_2 - \mathbf{p}_1)u^+(p_2)u(p_1) \qquad (12)$$

It is easy to see now that equation (12) is essentially a statement about the nonrenormalization of the constant for the corresponding conserved current J_μ^i at zero momentum transfer.

Let us take as an example the weak vector current in the usual form,[4,5] assuming the strong interactions to be invariant under $SU(3)$:

$$J_\mu^V(x) = \cos\theta\, J_\mu^{\Delta S=0}(x) + \sin\theta\, J_\mu^{\Delta S=1}(x) \qquad (13)$$

$$= \cos\theta(J_\mu^1 + iJ_\mu^2) + \sin\theta(J_\mu^4 + iJ_\mu^5) \qquad (14)$$

The vector current contribution to the amplitude for the leptonic decay

$$B \to B' + l + \nu$$

is given by

$$S = -\frac{iG_0}{\sqrt{2}} \int d^4x \langle B'(p')|J_\mu^V(x)|B(p)\rangle e^{iqx} \bar{u}_l(p_l)\gamma_\mu(1 + \gamma_5)u_\nu p(\nu) \qquad (15)$$

where $q = -(p_l + p_\nu)$ is the momentum transfer.

From translational and relativistic invariance, we can write (if, for instance, B and B' are spin-$\frac{1}{2}$ particles)

$$\langle B'(p')|J_\mu(x)|B(p)\rangle = e^{-i(p'-p)x} M_\mu(p, p') \qquad (16)$$

where

$$M_\mu(p, p') = \bar{u}(p')[A_1(q^2)\gamma_\mu + A_2(q^2)\sigma_{\mu\nu}q_\nu + A_3(q^2)q_\mu]u(p) \qquad (17)$$

In the zero momentum transfer limit, assuming the particle B at rest, we have

$$\lim_{p\to p'} M_\mu(p, p') = \delta_{\mu 4}A_1(0) \qquad (18)$$

Therefore, in the limit $q \to 0$ what will appear in (15) when the three-dimensional integration is performed is

$$\langle B'(p')| \int J_4^V d\mathbf{x} |B(p)\rangle$$

$$= \cos\theta\langle B'(p')|F^1 + iF^2|B(p)\rangle + \sin\theta\langle B'(p')|F^4 + iF^5|B(p)\rangle \qquad (19)$$

$$= u^+(p')\{\cos\theta(M_{B'B}^1 + iM_{B'B}^2) + \sin\theta(M_{B'B}^4 + iM_{B'B}^5)\}u(p)\delta(\mathbf{p}' - \mathbf{p}) \qquad (20)$$

It is clear from (20) that the effective coupling constant at zero momentum transfer is equal to the one we would have obtained had we calculated the matrix element of J_μ^V in (15) using free fields in the definition (6) for the currents. This is nothing more than the statement that at zero momentum transfer the weak vector coupling constant is not renormalized by the strong interactions.

The nonrenormalization of the charge is given if $F = Q$, and J_μ is the electromagnetic current. If the invariance group is $SU(2)$, the J_μ^i are the isotopic spin currents, and the F^i the total isotopic spin operators. This leads to the nonrenormalization of the strangeness-conserving weak vector coupling constant. In the case of $SU(3)$ considered here, the F^i are the eight unitary spins, and we are also led to the nonrenormalization of the strangeness-nonconserving vector coupling constant, etc. Note that if B and B' belong to the adjoint representation, the matrix elements $M_{B'B}^i$ are equal to the structure constants $f_{B'B}^i$ of the group.

Let us now assume that the violation of $SU(3)$ invariance can be described by a local Lagrangian $L_1(x)$ which transforms as an $I = 0$, $S = 0$ member of an $SU(3)$ multiplet (but not necessarily like the eighth component of an octet). We shall prove that the first-order correction to the vector decay amplitude in the limit of zero momentum transfer can be accounted for by using unrenormalized coupling constants, but with wave functions corresponding to the physical masses. As we shall see, this wave function correction is in fact only of second order for the relevant matrix element. So in first order there is no modification of the matrix element of the vector current.

The matrix element corresponding to the first-order correction is given by

$$S_{fi}^{(1)} = \frac{G_0}{\sqrt{2}} \int d^4x\, e^{iqx} M_\mu^{(1)}(x) \bar{u}_l \gamma^\mu (1 + \gamma_5) u_v \tag{21}$$

with

$$M_\mu^{(1)}(x) = -\langle B'(p')| \int T(J_\mu^V(x) L_1(x')) |B(p)\rangle d^4x' \tag{22}$$

In the limit $q \to 0$, $S_{fi}^{(1)}$ is proportional to

$$\int M_4^{(1)}(x) d^4x = -\int \langle B'|T[J_4^V(x) L_1(x')]|B\rangle d^4x\, d^4x' \tag{23}$$

$$= -\int d^4x'\, d^4x \{\sum_n \theta(t - t')\langle B'|J_4^V(x)|n\rangle\langle n|L_1(x')|B\rangle$$

$$+ \theta(t' - t)\langle B'|L_1(x')|n\rangle\langle n|J_4^V(x)|B\rangle\} \tag{24}$$

In performing first the integration over d^3x, there will appear the quantity $\int J_4^V(x) d\mathbf{x}$, which is nothing but a linear combination of the infinitesimal generators of the group. For reasons already explained, the only contribution in (24) will come from states $|n\rangle$ belonging to the same multiplet as $|B\rangle$ and $|B'\rangle$ and with an energy momentum close to that of $|B\rangle$ and $|B'\rangle$. Also, we can write

$$\langle n|L_1(x')|B\rangle \simeq \bar{\varphi}_B(x')\delta M_B\varphi_B(x') \tag{25}$$

where δM_B is the mass shift of particle B induced by the symmetry-breaking interaction L_1 in first order. To avoid divergences, it is convenient to describe the states $|B\rangle$ and $|B'\rangle$ by normalizable wave packets $\varphi_B(x)$ and $\varphi_{B'}(x)$, solutions of the Dirac equation with the degenerate mass M_0.

Finally, in the limit of zero momentum transfer, we can write

$$M_4^{(1)}(x) = -i\{\cos\theta(M_{B'B}^1 + iM_{B'B}^2) + \sin\theta(M_{B'B}^4 + iM_{B'B}^5)\}$$
$$\times \{\bar{\varphi}_{B'}(x)\gamma_4\delta\varphi_B(x) + \delta\bar{\varphi}_{B'}(x)\gamma_4\varphi_B(x)\} \tag{26}$$

where

$$\delta\varphi_B(x) = -i\int d^4x'\theta(t-t')[\sum_n \varphi_n(x)\bar{\varphi}_n(x')]\delta M_B\varphi_B(x') \tag{27}$$

$$\delta\bar{\varphi}_{B'}(x) = -i\int d^4x'\theta(t'-t)[\sum_n \bar{\varphi}_n(x')\varphi_n(x)]\delta M_B\bar{\varphi}_{B'}(x')$$

Let us compare $\delta\varphi_B(x)$ with the result of a first-order perturbation performed on the Dirac equation:

$$(\gamma_\mu\partial_\mu + M_0)\varphi'(x) = \delta M_B\varphi'(x) \tag{28}$$

the mass displacement term $\delta M_B\varphi'(x)$ being considered as the perturbation. It can be easily shown that we can identify $\varphi_B(x) + \delta\varphi_B(x)$ and $\bar{\varphi}_{B'}(x) + \delta\bar{\varphi}_{B'}(x)$ with the perturbed wave functions of the ingoing and outgoing baryons under the influence of the mass splitting δM_B and $\delta M_{B'}$. In other words, looking back at (26), the effect of the symmetry-breaking on the matrix element of the time component of the current can be accounted for by taking Dirac wave functions with the physical masses. More precisely, the matrix element of the time component of the current is proportional to the overlap between the function $\varphi_B(\mathbf{x}, t)$ and $\varphi_{B'}(\mathbf{x}, t)$ with the perturbed mass. By use of plane waves for ϕ_B and $\phi_{B'}$ and taking the particle B at rest, the lack of overlap is given by

$$1 - \sqrt{\frac{M_B + E'}{2E'}} \tag{29}$$

which is of the order of $(\delta M/M_0)^2$, in agreement with the known result of perturbation theory, where the lack of overlap between the unperturbed wave function and the first-order perturbed wave function is of second order. Therefore, to first order in the symmetry-breaking, there is no modification of the matrix element of the time component of the current.

In principle, our considerations apply as well to mesons as to baryons, except that in such case $\delta M/M_0$ is of order 1, and therefore the use of first-order perturbation is problematic unless we make some kind of dynamical assumption to explain the validity of the Gell-Mann–Okubo mass formula.

A remark should be made at this point concerning our assumption that the symmetry-breaking interaction can be described by a local Lagrangian $L_1(x)$. This limitation implies that our proof would not be valid, for instance, in a theory where the symmetry-breaking would be due to a second-order effect, as in certain subnuclear particle models where the symmetry-breaking term in the Largrangian transforms as the $I = 0$, $S = 0$ component of a unitary triplet.

In conclusion, I would like to stress that the proof I have given holds independently of the representation to which the symmetry-breaking interaction belongs, provided it carries the proper quantum numbers. Looking back to the proof, it is not even necessary that $L_1(x)$ have any particular transformation property under the group as long as B and B' belong to the same representation.

Ademollo and Gatto[1] have given a proof of the nonrenormalization of the vector coupling constants to first order in the symmetry-breaking interaction using different assumptions, namely, that the vector currents and the electromagnetic current belong to the same unitary octet and that the symmetry-breaking interaction transforms as the eighth component of an octet.

We shall give a slightly different proof† of Ademollo and Gatto's result, using the same assumptions as they do. For this purpose, it is useful to introduce a new frame through the rotation U belonging to $SU(3)$:[6]

$$U = e^{2i\theta F_7} \tag{30}$$

In this new frame (designated by primed quantities),

$$J_\mu^V = (J_\mu'^1 + iJ_\mu'^2) = J_\mu'(\Delta S' = 0) \tag{31}$$

where J_μ' corresponds to $\Delta S' = 0$, $|\Delta I'| = 1$.

With no symmetry-breaking, the leptonic decays involve the

---

†The proof quoted here is due to C. Bouchiat.

matrix elements

$$\langle B'|J'_\mu(\Delta S' = 0)|A'\rangle \tag{32}$$

with conservation of S'. The symmetry-breaking Lagrangian, on the other hand, transforms as

$$L_8 \sim r_0 = \alpha\eta'_0 + \beta\pi'_0 + \gamma K'_{01} \tag{33}$$

where α, β, and γ are well-defined constants which depend on θ.

To first order in the symmetry-breaking interaction, we have to consider a matrix element which transforms as

$$\langle A'|J'_\mu(\alpha\eta'_0 + \beta\pi'_0 + \gamma K'_{01})|B'\rangle \tag{34}$$

The term $J'_\mu K'_{01}$ will not contribute, since A' and B' have the same S'. To show that the term $J'_\mu \pi'_0$ is zero, let us consider the operation

$$G' = Ce^{i\pi I_i} = U^{-1}GU \tag{35}$$

For first-class matrix elements (those with γ_μ and $\sigma_{\mu\nu}q_\nu$), J'_μ is invariant under G'. As a result, $J'_\mu\pi'_0$ has the opposite G' parity and does not contribute to the matrix element. We are left now with the term $J'_\mu\eta'_0$. In the new frame, J'_μ is the isotopic spin current, which is known to be conserved even in the presence of the symmetry-breaking medium-strong interaction represented by η'_0. As a result, the vector coupling constant is not renormalized.

The proof of Ademollo and Gatto is not restricted to a Lagrangian theory, but it requires that the symmetry-breaking interaction transform as the eighth component of an octet. Furthermore, it cannot be extended in an obvious way to larger symmetry groups.

Recently, Fubini and Furlan[3] have reconsidered the problem of the partially conserved currents and have given a method which permits in principle the calculation of the higher-order corrections to the renormalized coupling constants. I shall only sketch the method, using as an example the matrix element involved in the decay

$$\bar{K}_0 \to \pi^+ + l^- + \gamma$$

that is,

$$G_0\langle\pi^+(p')|J_\mu^{\Delta S=1}(x)|\bar{K}_0(p)\rangle \tag{36}$$

$$= \frac{1}{(2\pi)^3}\frac{e^{i(p'-p)x}}{(4E_K E_\pi)^{1/2}}[(p+p')_\mu F_1(q^2) + (p'-p)_\mu F_2(q^2)] \tag{37}$$

with

$$q^2 = (p'-p)^2 \tag{38}$$

Let us consider as before

$$F_+ \equiv \int J_4^{\Delta S=1}(x)dx = F^4 + iF^5 \tag{39}$$

and

$$F_- = (F_+)^+ \tag{40}$$

for which we have the commutation relation

$$[F_+, F_-] = F_3 + \sqrt{3}\,F_8 \tag{41}$$

From (36), (37), and (39), we get

$$G_0\langle\pi^+(p')|F_+(t)|\bar{K}_0(p)\rangle = e^{i(E_K - E_\pi)t}\delta(\mathbf{p}' - \mathbf{p})G(p) \tag{42}$$

where

$$G(p) = \frac{1}{(4E_K E_\pi)^{1/2}}[(E_K + E_\pi)F_1(q_0^2) + (E_K - E_\pi)F_2(q_0^2)] \tag{43}$$

with

$$q_0^2 = (E_K - E_\pi)^2 = [(\mathbf{p}^2 + m_K^2)^{1/2} - (\mathbf{p}^2 + m_\pi^2)^{1/2}]^2 \tag{44}$$

Now consider (41) taken between physical π^+ states (with the F at $t = 0$). Let us insert a complete set of intermediate states $|\alpha\rangle$ which we separate into the $|\bar{K}_0\rangle$ state and all the others. Using equations (40) to (43), we get

$$G^2(p)\delta(\mathbf{p}' - \mathbf{p}) = G_0^2\delta(\mathbf{p}' - \mathbf{p}) + \delta G^2(p)\delta(\mathbf{p}' - \mathbf{p}) \tag{45}$$

where

$$\delta G^2(p)\delta(\mathbf{p}' - \mathbf{p}) = G_0^2\Big\{\sum_{\alpha \neq \bar{K}_0}\langle\pi^+(p')|F_+(o)|\alpha\rangle\langle\alpha|F_-(o)|\pi^+(p)\rangle$$

$$- \sum_\beta \langle\pi^+(p')|F_-(o)|\beta\rangle\langle\beta|F_+(o)|\pi^+(p)\rangle\Big\} \tag{46}$$

If there were no second term on the right side of equation (45), the effective coupling constant $G^2(p)$ would be unrenormalized and equal to G_0^2. We want to investigate more closely $\delta G^2(p)$.

Let us assume that only a part of the total Hamiltonian, call it H_0, is invariant under the symmetry group considered:

$$H = H_0 + fH_1 \tag{47}$$

We can write

$$\int\frac{\partial J_\mu^{\Delta S=1}}{\partial x_\mu}dx = \dot{F}_+(t) = -i[F_+^{(t)}, H] = -if[F_+^{(t)}, H_1] \tag{48}$$

Taking equation (48) between eigenstates of H, we get

$$\langle\beta|i\int\frac{\partial J_\mu^{\Delta S=1}(x)}{\partial x_\mu}dx|\alpha\rangle = (E_\alpha - E_\beta)\langle\beta|F_+(t)|\alpha\rangle \tag{49}$$

Defining

$$\frac{\partial J_\mu^{\Delta S=1}(x)}{\partial x_\mu} \equiv fD^{\Delta S=1}(x) \tag{50}$$

we get from (49) and (50)

$$\langle\beta|F_+(0)|\alpha\rangle = if\delta(\mathbf{p}_\alpha - \mathbf{p}_\beta)\frac{\langle\beta|D^{\Delta S=1}(0)|\alpha\rangle}{E_\alpha - E_\beta} \tag{51}$$

Substituting (51) in (46) we get finally

$$\delta G^2(p) = G_0^2 f^2\left\{\sum_{\alpha\neq\bar{K}_0}\frac{|\langle\pi^+(p)|D^{\Delta S=1}(0)|\alpha\rangle|^2}{(E_\pi - E_\alpha)^2}\delta(\mathbf{p}_\alpha - \mathbf{p})\right.$$
$$\left. -\sum_\beta\frac{|\langle\beta|D^{\Delta S=1}(0)|\pi^+(p)\rangle|^2}{(E_\pi - E_p)^2}\delta(\mathbf{p}_\beta - \mathbf{p})\right\} \tag{52}$$

It is clear from (52) that the renormalization of the coupling constant is only of second order in the symmetry-breaking Hamiltonian, as was found previously. Furthermore, it contains the damping factor $(E_\pi - E_\alpha)^2$ in the denominator. Also, the renormalization $\delta G^2(p)$ is a function of p. Fubini and Furlan give an argument according to which $\delta G^2(p)$ will be minimum for $|\mathbf{p}| \to \infty$. It is therefore advisable to define the renormalized coupling constant as $G_{K\pi} \equiv G(\infty) = F_1(0)$. In principle, the renormalization $\delta G^2(\infty)$ can be calculated from (48), (50), and (52) by introducing a dynamical model for the symmetry-breaking mechanism or using a dispersion-like approach to evaluate the matrix elements of the divergence of the current.

REFERENCES

1. M. Ademollo and R. Gatto, *Phys. Rev. Letters* **13**: 264 (1964).
2. C. Bouchiat and Ph. Meyer, *Nuovo Cimento* **34**: 1122 (1964).
3. S. Fubini and G. Furlan, *Physics* **1**: 229 (1965).
4. N. Cabibbo, *Phys. Rev. Letters* **10**: 531 (1962).
5. M. Gell-Mann, California Institute of Technology Report C.T.S.L. 20 1961; *Phys. Rev.* **125**: 1067 (1961).
6. B. d'Espagnat and J. Prentki, *Nuovo Cimento* **24**: 497 (1962); N. Cabibbo, *Phys. Rev. Letters* **12**: 62 (1964).

Multiplet Structure and Mass Sum Rules in $SU(6)$ Symmetry Scheme

VIRENDRA SINGH

TATA INSTITUTE OF FUNDAMENTAL RESEARCH
Bombay, India

Twenty-seven years ago, Wigner proposed the $SU(4)$ theory of nuclear forces.[1] It was suggested that the nuclear forces are independent not only of the orientation in isospin space but also of the ordinary-spin orientation of the nucleons. Thus, it does not matter whether the interacting particle is $p\uparrow$ or $p\downarrow$ or $n\uparrow$ or $n\downarrow$ ($\uparrow$ and $\downarrow$ refer respectively to the third component, "up" or "down," of the spin). This leads to the $SU(4)$ symmetry group of nuclear interaction, with the four $I = \frac{1}{2}$, $S = \frac{1}{2}$ nucleons belonging to the basic four-dimensional representation.[†] In this theory, one encounters for the first time a "marriage" of an internal symmetry group, in this case the $SU(2)$ isospin group, with the ordinary-spin $SU(2)$ group. Recently, Gursey, Radicati, and Pais[2] and, independently, Sakita[3] have proposed an $SU(6)$ theory for strongly interacting particles (hadrons) in which the basic ideas are quite similar to Wigner's $SU(4)$ theory. One again fuses the internal symmetry group for hadrons with the ordinary spin. The internal symmetry group is, of course, taken to be $SU(3)$, since this has emerged as the "right generalization" of the isospin $SU(2)$ group for hadrons.

I shall first discuss the multiplet assignments of various observed particles in the $SU(6)$ symmetry scheme, because the first thing one demands from a symmetry theory is that it predict cor-

---

[†]One could refer to these four basic objects in $SU(4)$ theory as a "Beatle Quartet."

rect multiplet structure. We shall see that $SU(6)$ is very satisfactory on this point. At this level of comparison with experiment, one neglects the mass differences among various particles in a multiplet. As one knows, however, these mass differences are not very "small," and one would like to have for masses in a multiplet some kind of sum rules that hold also in the presence of symmetry-breaking interactions. Gell-Mann–Okubo[4] (GM-O) mass sum rules provide this for $SU(3)$ symmetry. So we shall discuss this problem for the $SU(6)$ scheme after assignments of various multiplets.

1. MULTIPLET ASSIGNMENTS IN $SU(6)$

Because $SU(6)$ is not as familiar as the $SU(3)$ group, we shall illustrate the procedure by first deriving some low-dimensional useful representations for $SU(3)$. It should be noted that since $SU(3)$ is a purely internal symmetry group, the particles in the same $SU(3)$ multiplet have to have the same space-time properties, that is, same spin and same parity assignment. In the $SU(6)$ scheme, on the other hand, the spin is an essential part of the symmetry group. Parity operation is, however, outside the group. Thus, the particles in the same multiplet can only have the same parity. They could, however, have different values of spin.

Let us now get some practice for $SU(6)$ by working out a few examples in the $SU(3)$ case. The two basic representations "3" and "3*" for $SU(3)$ are three-dimensional and can be formally taken to consist of three particles called "quarks," $q = (p^0, n^0, \lambda^0)$ for "3" and their antiparticles $\bar{q} = (\bar{p}^0, \bar{n}^0, \bar{\lambda}^0)$ for "3*."[5] Here, (p^0, n^0) have isospin $I = \frac{1}{2}$ and hypercharge $Y = \frac{1}{3}$, and λ^0 is an isosinglet with $Y = -\frac{2}{3}$. Just as in the case of the isospin $SU(2)$ group, where one can get all the representations by taking the multiple outer products of the basic representation, which is an $I = \frac{1}{2}$ doublet, one can get all the $SU(3)$ representations by taking outer products of quark triplets. In the "eightfold-way" version of $SU(3)$, only the representations obtained by taking outer products of n quarks and m antiquarks with $n - m \equiv 0$ (mod. 3) are regarded as physically interesting.

Multiplying three quark triplets, we get a singlet state which is totally antisymmetric in the exchange of two quark triplets, two octet states with mixed symmetry, and one decuplet with total symmetry under such exchange ($a = p^0, n^0, \lambda^0$, etc.)

$$q_a q_b q_c = \boxed{\begin{smallmatrix} a \\ b \\ c \end{smallmatrix}} \oplus \boxed{\begin{smallmatrix} a & b \\ c \end{smallmatrix}} \oplus \boxed{\begin{smallmatrix} a & c \\ b \end{smallmatrix}} \oplus \boxed{\begin{smallmatrix} a & b & c \end{smallmatrix}}$$

$$3 \otimes 3 \otimes 3 = 1 \oplus 8 \oplus 8 \oplus 10 \tag{1}$$

The 1, 8, and 10 representations obtained by this multiplication are taken to be suitable for low-lying baryons. The known $J = \frac{1}{2}^+$ baryons (N, Λ, Σ, Ξ) do form an octet, while the $J = \frac{3}{2}^+$ baryons ($N^*, Y_1^*, \Xi^*, \Omega^-$) can be assigned to a decuplet, while Y_0^* (1405) is probably a unitary singlet. The isospin-hypercharge content of observed multiplets is in accord with that predicted by (1).

Let us next multiply one quark and one antiquark state:

$$q_a(\bar{q})^b = \tfrac{1}{3}\delta_a^b q_c(\bar{q})^c \oplus [q_a(\bar{q})^b - \tfrac{1}{3}\delta_a^b q_c(\bar{q})^c]$$

$$3 \otimes 3^* = 1 \oplus 8 \tag{2}$$

These are suitable for low-lying mesons. The observed 0^- mesons ($K, \eta, \pi, \bar{K}$) do form an octet with (I, Y) content given by (2). There are nine observed 1^- mesons ($K^*, \omega, \phi, \rho, \bar{K}^*$) which can be put in a singlet and an octet. Thus, the situation on this point is very satisfactory.

Let us look at $SU(6)$ now. The $SU(6)$ group of Gursey, Radicati, and Pais and Sakita, as mentioned before, is to be regarded as a similar kind of enlargement of internal symmetry group $SU(3)$ by spin group, as the $SU(4)$ of Wigner was of isospin $SU(2)$ group. One can regard it as the group of unimodular unitary transformations on the three spin one-half quarks, without prejudice against their physical existence. We shall label the quarks as follows:

$$p^0\uparrow \quad n^0\uparrow \quad \lambda^0\uparrow \quad p^0\downarrow \quad n^0\downarrow \quad \lambda^0\downarrow$$
$$1 \qquad 2 \qquad 3 \qquad 4 \qquad 5 \qquad 6$$

Let us work out some low-dimensional useful representations of $SU(6)$. From the above discussion of $SU(3)$, we would expect the outer product of three quark sextets to be the one which would be useful for low lying baryons and the outer product of a quark sextet with an antiquark sextet to be the one which would be useful for mesons. We get ($i = 1, 2, \ldots, 6$, etc.)

$$q_i q_j q_k = \boxed{\begin{array}{c} i \\ j \\ k \end{array}} \oplus \boxed{\begin{array}{cc} i & j \\ k \end{array}} \oplus \boxed{\begin{array}{cc} i & k \\ j \end{array}} \oplus \boxed{i\,|\,j\,|\,k}$$

$$6 \otimes 6 \otimes 6 = 20 \oplus 70 \oplus 70 \oplus 56 \tag{3}$$

and

$$q_i(\bar{q})^j = \tfrac{1}{6}\delta_i^j q_k(\bar{q})^k \oplus [q_i(\bar{q})^j - \tfrac{1}{6}\delta_i^j q_k(\bar{q})^k]$$
$$6 \otimes 6^* = 1 \oplus 35 \tag{4}$$

Since one knows the unitary spin–ordinary spin content of the basic sextet, one can work out the $SU(3)$, $SU(2)$ spin content of all these representations in a straightforward manner. The result is

$$1 = (1, 1)$$
$$35 = (8, 3) + (1, 3) + (8, 1)$$
$$20 = (8, 2) + (1, 4)$$
$$70 = (8, 4) + (10, 2) + (8, 2) + (1, 2)$$
$$56 = (10, 4) + (8, 2)$$

where the notation is (dimension of $SU(3)$ representation, spin multiplicity). It is very gratifying that the 0^- meson octet, the 1^- singlet, and the 1^- octet of observed mesons can all be accommodated in the 35-plet, while the baryon $\tfrac{1}{2}^+$ octet and $\tfrac{3}{2}^+$ decuplet are very naturally accommodated in the 56-plet. Also, a natural reason for vector mesons to occur as a nonet, while PS mesons occur as an octet, is provided.

We now come to the problem of mass sum rules in $SU(6)$.

2. MASS SUM RULES

The GM-O mass formula in $SU(3)$ is derived under the assumption that the symmetry-breaking mass operator transforms as an $I = Y = 0$ member of an $SU(3)$ octet. This gives for the mass $M(d, I, Y)$ of (I, Y) member of $SU(3)$ multiplet d,

$$M(d, I, Y) = M_0(d) + a(d)Y + b(d)[I(I + 1) - (Y^2/4)] \tag{5}$$

The expression (5), for octets, leads to

$$3M(8, 0, 0) + M(8, 1, 0) = 2M(8, \tfrac{1}{2}, 1) + 2M(8, \tfrac{1}{2}, -1) \tag{6}$$

and, for decuplets, to

$$M(10, \tfrac{3}{2}, 1) - M(10, 1, 0) = M(10, 1, 0) - M(10, \tfrac{1}{2}, -1)$$
$$= M(10, \tfrac{1}{2}, -1) - M(10, 0, -2) \quad (7)$$

The application of (6) to the baryon $J = \tfrac{1}{2}^+$ octet, and to 0^- mesons is very satisfactory. While an application of (6) to 1^- mesons is numerically not that satisfactory, the discrepancy is not much and is explainable in terms of "repulsion" between $I = Y = 0$ member of a 1^- octet and a neighboring 1^- singlet level. The $J = \tfrac{3}{2}^+$ baryon decuplet satisfies (7) very accurately.†

Thus, we can regard the $SU(3)$ octet nature of symmetry-breaking as established. We must therefore require the mass operator in the $SU(6)$ scheme also to transform as a linear combination of $I = Y = 0$ members of an $SU(3)$ singlet and octet. It should, further, be a scalar under spin transformations, as the symmetry-breaking term should respect angular momentum conservation. Only certain $SU(6)$ representations will have the components having desired transformation properties for the total mass operator, their dimensions are

$$1, 35, 189, 405, \ldots \quad (8)$$

They correspond, respectively, to the traceless $SU(6)$ tensors

$$T, \qquad T^{a}_{\alpha}, \qquad T^{\{a,b\}}_{\{\alpha,\beta\}}, \qquad T^{[a,b]}_{[\alpha,\beta]}, \qquad \ldots$$

where $\{\ \}$ and $[\ ]$ in tensor indices stand for antisymmetry and symmetry, respectively, under exchange of two symbols inside the brackets. The dimensionalities of these tensors are easy to calculate, and in a derivation explaining notations are

$$1 = 1$$
$$35 = 6 \times 6 - 1$$
$$189 = \left[\frac{6 \times (6-1)}{2} \right]^2 - 6 \times 6$$

---

†Further evidence for the octet nature of symmetry breaking in $SU(3)$ is provided by sum rules obeyed by decay widths Γ_α of various decuplet $J = \tfrac{3}{2}^+$ members into a $J = \tfrac{1}{2}^+$ baryon and 0^- mesons. We have

$$\Gamma_\alpha = |G_\alpha|^2 p^3 (m_B/m_D)$$

where p is the decay momentum and m_B and m_D are relevant $J = \tfrac{1}{2}^+$ baryon and $J = \tfrac{3}{2}^+$ decuplet masses, respectively. One of the sum rules is[6]

$$(2)^{1/2} G(N^*, N\pi) - 2G(\Xi^*, \Xi\pi) = 3G(Y_1^*, \Lambda\pi) - (3/2)^{1/2} G(Y_1^*, \Sigma\pi)$$

By putting in the experimental numbers, the left-hand side works out to be 7.58 ± 0.83 BeV^{-1}, while the right-hand side is 7.44 ± 0.67 BeV^{-1}.

$$405 = \left[\frac{6 \times (6 + 1)}{2}\right]^2 - 6 \times 6, \text{ etc.}$$

The numbers with the minus signs in these expressions represent the number of tracelessness conditions.

Before proceeding further, let us see which of the representations in series (8) will contribute to the mass differences for 35-plets, 56-plets, and 70-plets. Toward this end the following results are relevant:

$$35 \otimes 35 = 1 \oplus 35_s \oplus 35_a \oplus 189 \oplus 280 \oplus 280^* \oplus 405$$
$$56 \otimes 56^* = 1 \oplus 35 \oplus 405 \oplus 2695$$
$$70 \otimes 70^* = 1 \oplus 35_s \oplus 35_a \oplus 189 \oplus 280 \oplus 280^* \oplus 405 \oplus 3675$$

We thus see that only the 35, 189, and 405 members of series (8) will contribute for a meson 35-plet, while for a baryon 56-plet only 35, 405, and 2695 are relevant. For a 70-dimensional representation, the contribution to mass splittings can only come from 35-, 189-, 405-, and 3675-dimensional $SU(6)$ tensors.[†] We shall not consider the contributions to the mass splittings from all $SU(6)$ representations; only those from 35, 189, and 405.

Symmetry Breaking Like 35-Plet[7]

In this case the $I = 0$, $Y = 0$, $J = 0$, $SU(3)$ octet member is given by $(-T_3^3 \quad -T_6^6)$. From Ginibre's theorem,[8]

$$T_\nu^\mu \sim a_1 A_\nu^\mu + a_2 (A \cdot A)_\nu^\mu + \cdots + a_5 (A \cdot A \cdot A \cdot A \cdot A)_\nu^\mu \quad (9)$$

where A_ν^μ are 35 generators of $SU(6)$ ($\sum A_\lambda^\lambda = 0$). If one is interested only in the matrix elements $\langle D | T_\nu^\mu | D \rangle$, where $D \otimes D^*$ contains a 35-plet twice at most, then one can reduce (9) to

$$T_\nu^\mu \sim a_1' A_\nu^\mu + a_2' (A \cdot A)_\nu^\mu \quad (10)$$

where $a_1, \ldots, a_5$ and a_1', a_2' depend only on the Casimir operators of the $SU(6)$ group. Using expression (10), with $\mu = \nu = 3, 6$, one is led to the mass operator

$$M_{(35)}^{(8)} = a_1 + b_1 Y + c_1 [2S(S + 1) - C_2^{(4)} + (Y^2/4)] \quad (11)$$

Here the notation is $M_{(D)}^{(d)}$ for the "irreducible" mass operators trans-

---

[†]The interest in a 70-dimensional representation comes from the fact that it occurs in $35 \otimes 56 = 56 \oplus 70 \oplus 700 \oplus 1134$. Thus the 70-plet is a candidate for baryon resonances. As the content of the 70-plet includes a $J = \frac{3}{2}$, $SU(3)$ octet, the observed $D_{3/2}$ baryon resonances presumably belong to it.

forming like a linear combination of an $SU(6)$ singlet and an $I = Y = J = 0$ member of an $SU(3)$ d-plet belonging to an $SU(6)$ D-plet. Here S is the spin of λ^0-type quarks,

$$S_+ = A_3^6, \qquad S_- = A_6^3, \qquad S_3 = \tfrac{1}{2}(A_3^3 - A_6^6) \tag{12}$$

and $C_2^{(4)}$ is the second-order Casimir operator of the $SU(4)$ group generated by p^0 and n^0-type quarks, i. e.,

$$C_2^{(4)} = \tfrac{1}{2} \sum_{\lambda,\mu=1,2,4,5} \{A_\lambda^\mu, A_\mu^\lambda\} \tag{13}$$

Further, for a 56-plet (11) reduces (since in this case $56 \otimes 56^*$ contains 35 only once) to

$$M_{(35)}^{(8)} = a_1' + b_1' Y \tag{14}$$

Thus, we see that $M_{(35)}^{(8)}$ does not lift the J-degeneracy of $SU(6)$ at all. Thus, K^* and K, ρ and π, N^* and N, etc. remain degenerate. Further there is no Λ, Σ splitting for the baryon octet. Thus $M_{(35)}^{(8)}$ is not satisfactory and one has to consider the mass operators $M_{(189)}^{(1)}$, $M_{(189)}^{(8)}$, $M_{(405)}^{(1)}$, and $M_{(405)}^{(8)}$ also.

Symmetry Breaking Like 189- and 405-Plet[9]

We need the analog of expression (9) for $T_{[a,b]}^{[\alpha,\beta]}$ and $T_{\{a,b\}}^{\{\alpha,\beta\}}$ for this purpose. It is easy to find this, however, and since we shall be interested only in $SU(6)$ representations D, for which $D \otimes D^*$ contains 189 and 405 only once, it is unnecessary to give these general constructions. All one needs is the analog of expression (10), which, e.g., for $T_{\{a,b\}}^{\{\alpha,\beta\}}$ is given by

$$T_{\{a,b\}}^{\{\alpha,\beta\}} \sim C[\{A_a^\alpha, A_b^\beta\} + \{A_a^\beta, A_b^\alpha\} - \tfrac{1}{8}\delta_a^\alpha \sum_\lambda \{A_b^\lambda, A_\lambda^\beta\}$$

$$-\tfrac{1}{8}\delta_b^\alpha \sum_\lambda \{A_a^\lambda, A_\lambda^\beta\} - \tfrac{1}{8}\delta_a^\beta \sum_\lambda \{A_b^\lambda, A_\lambda^\alpha\}$$

$$-\tfrac{1}{8}\delta_b^\beta \sum_\lambda \{A_a^\lambda, A_\lambda^\alpha\} + \tfrac{1}{28}(\delta_a^\alpha \delta_b^\beta + \delta_b^\alpha \delta_a^\beta)C_2^{(6)}]$$

and a similar expression for $T_{[a,b]}^{[\alpha,\beta]}$. Here $C_2^{(6)}$ is the Casimir operator of $SU(6)$ group

$$C_2^{(6)} = \tfrac{1}{2} \sum_{\lambda,\mu} \{A_\lambda^\mu, A_\mu^\lambda\} \tag{16}$$

Now our task is to find the components of 189- and 405-rank tensors, which have $J = I = Y = 0$ and are $SU(3)$ singlets or octets. Let me illustrate the procedure by finding $J = I = Y = 0$, $SU(3)$ singlet member of 405-dimensional $SU(6)$ tensor.

We first find states and determine their quantum numbers for a 21-plet by extracting the symmetric part of $6 \otimes 6$. (The antisymmetric part gives a 15-plet. We have $6 \otimes 6 = 21 \oplus 15$.) We get in 21 a $J = 1$, $SU(3)$ sextet and a $J = 0$, 3^*-plet. Let the state in the 21-plet with quantum numbers I, I_3, Y, J, and J_3 and belonging to an $SU(3)$ d-plet be denoted by $\chi^{(d)}_{II_3YJJ_3}$. Thus, for example,

$$\chi^{(6)}_{0,0,-4/3,1,1} = T_{3,3}$$

$$\chi^{(6)}_{0,0,-4/3,1,0} = \frac{1}{\sqrt{2}}(T_{3,6} + T_{6,3}) \tag{17}$$

$$\chi^{(6)}_{0,0,-4/3,1,-1} = T_{6,6}$$

Since we are interested in only $I = Y = J = 0$, we form the following sum:

$$\chi^{(d)}_{I,Y;J} = \sum_{I_3,J_3} \overline{(\chi^{(d)}_{II_3YJJ_3})}(\chi^{(d)}_{II_3YJJ_3}) \tag{18}$$

Thus, for example,

$$\chi^{(6)}_{0,-4/3;1} = \tfrac{1}{2}T^{[3,3]}_{[3,3]} + T^{[3,6]}_{[3,6]} + \tfrac{1}{2}T^{[6,6]}_{[6,6]} \tag{19}$$

Now one knows that in $21 \otimes 21^* = 1 \oplus 35 \oplus 405$, there are only two possible $SU(3)$ singlets, one of which occurs in an $SU(6)$ singlet and another in an $SU(6)$ 405-plet. The $SU(6)$ singlet $\Phi^{(1)}_{(1)}$ is given by

$$\Phi^{(1)}_{(1)} = [\chi^{(6)}_{1,2/3;1} + \chi^{(6)}_{1/2,-1/3;1} + \chi^{(6)}_{0,-4/3;1}] + [\chi^{(3^*)}_{1/2,-1/3;0} + \chi^{(3^*)}_{0,2/3;0}] \tag{20}$$

The $SU(6)$ 405-plet and the $SU(3)$ singlet $\Phi^{(1)}_{(405)}$ must be a linear combination of two possible $SU(3)$ singlets that one can form out of the $SU(3)$ 6-plet and 3^*-plet which occur in the $SU(6)$ 21-plet and their antiparticles:

$$\Phi^{(1)}_{(405)} = [\chi^{(6)}_{1,2/3;1} + \chi^{(6)}_{1/2,-1/3;1} + \chi^{(6)}_{0,-4/3;1}] + \alpha[\chi^{(3^*)}_{1/2,-1/3;0} + \chi^{(3^*)}_{0,2/3;0}] \tag{21}$$

The orthogonality of $\Phi^{(1)}_{(1)}$ with $\Phi^{(1)}_{(405)}$ fixes the undetermined coefficient α, and we get

$$\alpha = -6 \tag{22}$$

A similar procedure is followed for other possible required components.

Using expressions (15) and (21), we are led to

$$M^{(1)}_{(405)} = a_4 + b_4[2J(J + 1) + C^{(3)}_2] \tag{23}$$

where $C^{(3)}_2$ is second-order Casimir operator for the $SU(3)$ group inside the $SU(6)$ group. Similarly, one obtains

$$M^{(1)}_{(189)} = a_2 + b_2[2J(J+1) - C_2^{(3)}] \tag{24}$$

$$M^{(8)}_{(189)} = a_3 + b_3\Big[\{2J(J+1) - C_2^{(3)}\}$$
$$+ 3\Big\{2I(I+1) - \frac{Y^2}{2} - 2N(N+1) + 2S(S+1)\Big\}$$
$$- \tfrac{3}{4}\Big\{2S(S+1) - C_2^{(4)} + \frac{Y^2}{4}\Big\}\Big] \tag{25}$$

$$M^{(8)}_{(405)} = a_5 + b_5\Big[\{2J(J+1) + C_2^{(3)}\}$$
$$+ \tfrac{21}{8}\Big\{2S(S+1) - C_2^{(4)} + \frac{Y^2}{4}\Big\}$$
$$+ 3\Big\{2I(I+1) - \frac{Y^2}{2} + 2N(N+1) - 2S(S+1)\Big\}\Big] \tag{26}$$

Here N is the spin of non-λ^0-type quarks.

If the mass operator contains all the above contributions, then we get for the mass operator M, using (11), (23), (24), (25), and (26),

$$M = a + bC_2^{(3)} + cJ(J+1) + dY + e\Big[2S(S+1) - C_2^{(4)} + \frac{Y^2}{4}\Big]$$
$$+ f[N(N+1) - S(S+1)] + g\Big[I(I+1) - \frac{Y^2}{4}\Big] \tag{27}$$

Applications of the Mass Operator Equation (27)

A. For the 56-dimensional representation of $SU(6)$, we have the following identities:[9]

$$2J(J+1) - C_2^{(3)} = -\tfrac{9}{2} \tag{28}$$

$$2S(S+1) - C_2^{(4)} + \frac{Y^2}{4} = -8Y - \tfrac{15}{2} \tag{29}$$

$$I(I+1) - \frac{Y^2}{4} - N(N+1) + S(S+1) = -Y + \tfrac{3}{4} \tag{30}$$

so the mass operator collapses into[9]

$$M = M_0 + M_1 J(J+1) + M_2 Y + M_3\Big[I(I+1) - \frac{Y^2}{4}\Big] \tag{31}$$

Thus, all the eight independent baryon masses of N, Λ, Σ, Ξ, N^*, Y_1^*, Ξ^*, and Ω^- are given in terms of four parameters. Comparing with expression (5), we see that prediction of our baryon mass operator is

$$a(8) = a(10) = M_2$$

$$b(8) = b(10) = M_3 \qquad (33)$$

The numerical accuracy to which (31) is satisfied is very good.

B. Application to a 35-plet of mesons. In this case we only get one sum rule, since our mass operator is much too general,

$$4K = 3\eta + \pi \qquad (34)$$

[meson labels = (meson mass)^2], which is GM-O sum rule for a 0^- meson octet.

To get any other sum rule we have to make extra assumptions, for example, $M_{(189)}^{(8)} = 0$; that is, $f = g$ leads to[9]

$$\omega\phi = \tfrac{1}{2}(\pi + K^* - K)(3K^* - \rho + K - \pi)$$
$$- \tfrac{1}{6}[(4K^* - \rho)(5K^* - \rho + \pi - K - 2\omega - 2\phi)] \qquad (35)$$

If we further put $f = g = 0$, we get an extra sum rule[11]

$$K^* - K = \rho - \pi \qquad (36)$$

and (35) collapses into†

$$(\omega - \rho)(\phi - \rho) = \tfrac{4}{3}(K^* - \rho)(\omega + \phi - 2K^*) \qquad (37)$$

Rules (35), (36), and (37) are all numerically very well satisfied with observed meson masses.

C. Mass operator (27) has also been used for the 70-plet with $f = g$, and seven sum rules have been obtained.[12] These sum rules have been used by Gyuk and Tuan in connection with tentative assignments of baryon resonances to the 70-plet.[13] It is probably of interest to give the mass sum rules with the assumption $b = f = 0$ also. One then gets eight sum rules, and the mass operator is diagonal in the U-chain. The sum rules, in the notation of reference 12, are

$$3\Lambda_\gamma + \Sigma_\gamma = 2N_\gamma + 2\Xi_\gamma \qquad (38)$$

$$\Sigma_\gamma - \tilde{\Sigma}_U = \Lambda_\gamma - \tilde{\Lambda}_U = \Xi_\gamma - \tilde{\Xi}_U^* \qquad (39)$$

$$\tilde{N}^* - \tilde{\Omega} = 3(\tilde{Y}_U^* - \tilde{\Xi}_U^*) = 3(\tilde{\Sigma}_U - \tilde{\Xi}_U) \qquad (40)$$

$$\tilde{N}^* + \tilde{\Omega} = \tilde{Y}_U^* + \tilde{\Xi}_U \qquad (41)$$

---

†Sum rules (36) and (37) were also obtained by Schwinger[10] by a different approach.

$$2\widetilde{N} + 2\widetilde{\Xi}_U = 3\Lambda_U + \widetilde{Y}_U^* \tag{42}$$

$$2\widetilde{N} + 2\widetilde{\Xi}_U^* = 3\widetilde{\Lambda}_U + \widetilde{\Sigma}_U \tag{43}$$

D. Dyson and Xuong have applied the mass operator (27) to $Y = 2$ states of 490-dimensional $SU(6)$ representation for two baryon states.[14] In this case, the mass operator collapses into

$$M = A' + B'I(I + 1) + B''J(J + 1) \tag{44}$$

One then has three sum rules (using particle names, in the notation of reference 14, for their masses)

$$3(D_{03} + D_{30}) + 2(D_{01} + D_{10}) = 5(D_{12} + D_{21}) \tag{45}$$

$$D_{30} - D_{03} = 3(D_{21} - D_{12}) = 6(D_{10} - D_{01}) \tag{46}$$

Experimentally,

$$D_{10} \approx D_{01}$$

REFERENCES

1. E.P. Wigner, *Phys. Rev.* **51**: 105 (1937).
2. F. Gursey and L. Radicati, *Phys. Rev. Letters* **13**: 173 (1964); A. Pais, *Phys. Rev. Letters* **13**: 175 (1964); F. Gursey, A. Pais, and L. Radicati, *Phys. Rev. Letters* **13**: 299 (1964).
3. B. Sakita, *Phys. Rev.* **136B**: 1756 (1964).
4. M. Gell-Mann, *Phys. Rev.* **125**: 1067 (1962) and CTSL-20 (1961); S. Okubo, *Progr. Theoret. Phys.* **27**: 949 (1962).
5. M. Gell-Mann, *Phys. Letters* **8**: 214 (1964); G. Zweig, CERN preprint.
6. V. Gupta and V. Singh, *Phys. Rev.* **135B**: 1442 (1964).
7. T.K. Kuo and T. Yao, *Phys. Rev. Letters* **13**: 418 (1964).
8. J. Ginibre, *J. Math. Phys.* **4**: 720 (1963).
9. Mirza A.B. Beg and V. Singh, *Phys. Rev. Letters* **13**: 418 (1964) and "Erratum," *Phys. Rev. Letters* **13**: (1964).
10. J. Schwinger, *Phys. Rev.* **135B**: 816 (1964).
11. P. Babu, *Nuovo Cimento* **33**: X, 654 (1964).
12. Mirza A.B. Beg and V. Singh, *Phys. Rev. Letters* **13**: 509 (1964).
13. I.P. Gyuk and S.F. Tuan, *Phys. Rev. Letters* **14**: 121 (1965).
14. F.J. Dyson and N.H. Xuong, *Phys. Rev. Letters* **13**: 815 (1964) and "Erratum," *Phys. Rev. Letters* **14**: 339 (1965).

A New Approach to Scattering Theory

R. BLANKENBECLER

PRINCETON UNIVERSITY
Princeton, New Jersey

For a review of the appropriate reference, and for more precise definitions of the quantities used here, I refer you to a recent article by R. Sugar and myself.[1] Our purpose is to develop upper and lower bounds for the phase shifts and the K-matrix elements of complicated problems. We will call a problem solved if we can write it in such a form that it can be easily integrated on a computer. All three-body and higher scattering problems are unsolvable in the sense that they cannot be coded for a finite-priced computer.

In pursuing our goal, there are actually two distinct steps involved. Before we invent upper and lower bounds for the K-matrix elements, we must first prove that a solution actually exists. There are numerous papers on this subject, but the one which I consider the simplest and most direct, and certainly the closest in spirit to our present discussions, is given in reference 1. We will ignore this part of the problem here and get on with the physics.

We will first discuss single-channel scattering in order to introduce the type of manipulations to be used in more complicated problems. Then the multichannel (or multiparticle) case will be thoroughly considered in order to get upper and lower bounds on the elements of the K-matrix. Finally, a few remarks about the application of our methods to the relativistic case will be given.

1. SINGLE-CHANNEL SCATTERING

Our first project is to find out what happens to the phase shift as we change the potential. The potential will be assumed to be a function of a parameter x. Then the phase shift is given by

$$k \tan \delta(x) = -\langle l|VR|l \rangle \tag{1}$$

where

$$R = [1 - GV]^{-1}$$

The standing wave Green's function is denoted by G, and $|l\rangle$ is the free spherical wave with angular momentum l. By differentiating, we find

$$k \frac{\partial}{\partial x} \delta(x) = -\left\langle x \left| \frac{\partial V}{\partial x} \right| x \right\rangle$$

where

$$|x\rangle = R|l\rangle \cos \delta(x)$$

This is just the continuum version of the Feynman–Hellmann theorem on the parametric change of eigenvalues. It is not very difficult to integrate this equation, and we get

$$k[\delta(1) - \delta(0)] = -\int_0^1 dx \left\langle x \left| \frac{\partial V}{\partial x} \right| x \right\rangle \tag{2}$$

This tells us that if the derivative of V has a fixed sign, then we know that the phase shift changes in a definite direction. Therefore, if we write

$$V(x) = V_0 + x(V - V_0)$$

and if

$$(V - V_0) \leqslant 0$$

then we can get a lower bound on $\delta(1) = \delta(V)$ by solving $\delta(0)$. This is fine if we can construct a solvable V_0 which is very close to V. This is easy to do by using Schwartz's inequality.

Let us assume that V is negative-definite; then we can write that

$$V \leqslant V_0 = V_s = V|q\rangle\langle q|V|q\rangle^{-1}\langle q|V \tag{3}$$

for any state $|q\rangle$. This is a separable potential and can be easily solved:

$$k \tan \delta(0) = -\langle l|V|q\rangle^2 (\langle q|V - VGV|q\rangle)^{-1} \qquad (4)$$

and we know that $\delta(V) \geqslant \delta(0)$.

The condition for a resonance at E is

$$\langle q|V - VG(E)V|q\rangle = 0$$

Since the phase shift we have calculated is a lower bound on the true phase shift, we know that E is larger than the true resonance position.

It is easy to show that the phase shift given by equation (4) is exact if $|q\rangle$ is the exact wave function. In fact, this expression forms a stationary principle for the phase shift and is the same as the Schwinger variational principle. We know that it also provides a bound which is very important in numerical applications.

To get the other bound, we must find a solvable potential U which satisfies

$$U \leqslant V$$

and write

$$V \equiv U + W \qquad (5)$$

where W is positive-definite. Now if we make W separable in the same manner as in equation (3), we can write

$$V(x) = U + W_S + x(W - W_S)$$

Since the derivative of V is now positive-definite, we know that

$$\delta(V) \leqslant \delta(U + W_S) \qquad (6)$$

This is now a stationary upper bound and requires a knowledge of the phase shift in the potential U and the associated Green's function.

2. COUPLED CHANNELS

In this case, the Schrödinger equation is a matrix equation and the wave function is a column vector with a separate component for each channel. The potential matrix is written as

$$V = \begin{pmatrix} V_{NN} & V_{NM} \\ V_{MN} & V_{MM} \end{pmatrix}$$

where V_{NN} is an $N \times N$ matrix, V_{NM} is $N \times M$, and so on. The lower M components, which we assume to be unexcited and therefore to have thresholds above the energy range of interest, can be explicitly eliminated and we find an $N \times N$ problem with the effective potential

$$W = V_{NN} + V_{NM}(E - H_{MM})^{-1}V_{MN} \tag{7}$$

As in the single-channel case, our problem will be to find good upper and lower variational bounds for W so that we can calculate upper and lower bounds for elements of the K-matrix.

The K-matrix is conveniently introduced as

$$K = -k^{-1/2}W[1 - GW]^{-1}k^{-1/2}$$

where G is a diagonal standing wave Green's function matrix and k is a diagonal phase space matrix made up of the components of momentum in the open N channels.

If W depends on a parameter x, then we find

$$\langle\phi|K(1) - K(0)|\phi\rangle = -\int_0^1 dx\langle x|\partial W/\partial x|x\rangle \tag{8}$$

where

$$|x\rangle \equiv [1 - GW]^{-1}k^{-1/2}|\phi\rangle$$

One must use this equation with some care, since the standing wave state $|x\rangle$ can have singularities for values of x between zero and one. These simply mean that one must keep track of the branch of the phase shift. In practice, there is no difficulty arising from this point.

Our previous arguments allow us to construct upper and lower bounds on any diagonal element of K by constructing upper and lower bounds for W. If we have two operators W_u and W_l such that

$$W_u \leqslant W \leqslant W_l$$

then by letting ϕ contain only the free spherical wave with component i, we find

$$K_{ii}(l) \leqslant K_{ii} \leqslant K_{ii}(u) \tag{9}$$

in the sense of the phase. By letting ϕ contain two components i and j, we find

$$D(l) \leqslant D \leqslant D(u)$$

where

$$D \equiv K_{ii} + K_{jj} + 2K_{ij}$$

Using our upper and lower bounds on the diagonal elements, we find

$$K_{ij}(l) - L \leqslant K_{ij} \leqslant K_{ij}(u) + L \tag{10}$$

where
$$L = K_{ti}(u) + K_{jj}(u) - K_{ti}(l) - K_{jj}(l)$$

Therefore, all we have to do is to produce explicitly the operators $W_{u,l}$, which are bounds and which lead to solvable problems.

We define
$$W_l = V_{NN} + V_{NM}|q\rangle\langle q|E - H_{MM}|q\rangle^{-1}\langle q|V_{MN} \tag{11}$$

where $|q\rangle$ is an arbitrary state with M components, and also write
$$W(x) = W_l + x W_{NM} B^{-1}(x) V_{MN}$$

where
$$B(x) = E - T_{MM} - x V_{MM} + \frac{x}{1-x}(E - H_{MM})|q\rangle\langle q|(E - H_{MM})$$

It is easy to show that $W(1) = W$ and that the derivative of W is negative for E *below* the lowest threshold among the M channels. Note that our proof does not depend on the absence (or presence) of bound states of the operator H_{MM}.

To get the upper bound, we need to know a lower bound to the spectrum of H_{MM}. For a discussion of a procedure for actually calculating such a bound, see reference 1. The lower bound is called E_M, and we have
$$h \equiv H_{MM} - E_M \geqslant 0 \tag{12}$$

Then by making this positive operator separable, we find
$$E - H_{MM} = E - E_M - h \leqslant E - E_M - h_s$$

Now the one-parameter effective potenial is chosen to be
$$W(x) = V_{NM} + V_{NM} A^{-1}(x) V_{MN} \tag{13}$$

where
$$A(x) = E - E_N - h_S - x(h - h_S)$$

It is easily seen that the derivative of W is positive definite, and therefore we may chose
$$W_u = W(0) \tag{14}$$

The operator $A^{-1}(0)$ is easily calculated since it involves a separable kernel h_S.

3. EXAMPLE

In order to explore our methods a bit, let us look at the situation in which a bound state occurs in H_{MM}. The lower bound case will be sufficient for our purposes. We will assume that our trial function is sufficiently accurate that if used as a Rayleigh–Ritz trial function, a bound state occurs:

$$\langle q|H_{MM}|q\rangle = \epsilon \geqslant \epsilon_{\text{true}}$$

For energies in the neighborhood of ϵ, we may neglect V_{NN} (or include it, as you wish) and find

$$W_l = V_{NM}|q\rangle(E-\epsilon)^{-1}\langle q|V_{MN} \qquad (15)$$

The K-matrix arising from this separable potential is easy to calculate

$$k^{1/2}Kk^{1/2} \equiv -V_{NM}|q > (E-\epsilon-\langle q|V_{MN}GV_{NM}|q\rangle)\langle q|V_{MN} \quad (16)$$

Thus we see that K has a resonance at the point $\epsilon + \langle q|VGV|q\rangle$, where the second term is the level shift due to the coupling of the N and M sectors. Since ϵ is greater than the exact binding energy, we see that the exact phase shift will be larger than δ_l, as expected.

4. RELATIVISTIC CONSIDERATIONS

In the last part of my discussion, I would like to give a brief introduction to some work by R. Sugar and myself on extending the previous discussion to the relativistic case. In many ways, our final equations will look like the quasi-potential approach of Logunov *et al.* However, our potential has different analytic properties and we can handle multiparticle states. Also, the potential we will introduce can be constructed in a nonperturbative fashion.

Our starting point here will be the classic Bethe–Salpeter equation in the ladder approximation. After writing this equation in our desired form, we will find that the three-particle states are incorrectly handled. The B-S equation can always be fixed up, but it seems simpler just to take our equations as they stand and solve them.

The B-S equation will be written as

$$M(p,q) = V(p,q) - i\int [d^4k/(2\pi)^4]V(p,k)G_1G_2M(k,q) \qquad (17)$$

where

$$V = g^2[m^2 + (p - q)^2]^{-1}$$

and

$$G_{1,2} = [1 + (\tfrac{1}{2}p \pm k)^2]^{-1}$$

The difficulty with the B-S equation for our purposes (aside from the fact that it is almost impossible to solve) is that the Feynman propagators allow a given graph to contribute to several intermediate states. This will not allow us to apply our upper and lower bound technique, which requires that the different channels be separated.

In order to accomplish this goal, we must introduce new types of propagators. To separate off the two-particle singularities, it is essential that we introduce a propagator which is singular on both sides of the ladder if it is singular on one side. Therefore, we are led to define in the center of mass system, $P = (\mathbf{0}, s^{1/2})$,

$$E_2 = 2\pi \int ds'(s' - s)^{-1}\delta[1 + (\tfrac{1}{2}P' + k)^2]\delta(1 + (\tfrac{1}{2}P' - k)^2] \quad (18)$$

where

$$P' = (\mathbf{0}, s'^{1/2})$$

The integration can be carried out with the result that

$$E_2 = (\pi/2)\delta(k_0)[(k^2 + 1)^{1/2}(k^2 - q^2)]^{-1} \quad (19)$$

where

$$q^2 + 1 = s/4$$

Now we introduce another propagator R_2 by

$$-iG_1G_2 = E_2 + R_2$$

with the knowledge that R_2 does not lead to any two-particle singularities.

It is then simple to show that the B-S equation can be written as

$$M = W + WE_2M \quad (20)$$

where

$$W = V + VR_2(1 + VR_2)^{-1}V \quad (21)$$

The function W is the effective potential, or optical potential, which correctly introduces the multiparticle singularities into the solution of the two-particle equation (20).

Since E_2 [see equation (19)] contains a delta-function which forces k_0 to vanish, the integral equation (20) is a three-dimensional equation, but the intermediate integrations are over a relativistic phase space. This fact leads to an M which satisfies the correct relativistic unitarity relation in the elastic region.

The next step is to extract the three-particle singularities. This is done by defining a propagator E_3 which has three delta-functions under the s' integral. Then by defining

$$R_2 = E_3 + R_3$$

one can proceed as before. However, one finds that the ladder B-S equation is inconsistent in the three-particle sector. This is not too surprising, and it is due to the fact that once you allow a three-particle state, you must allow the third particle to be absorbed by either of the other two. This will immediately lead to bubble-type graphs on the legs of the ladder (in fact, any number of bubbles between rungs of the ladder), and this type of graph is not present in the ladder approximation. It seems simpler at this point to drop the crutch supplied by the B-S equation and work exclusively with the multiparticle equations, which are easily seen to be of the form

$$M_{22} = V_{22}(1 + E_2 M_{22}) + V_{23} E_3 M_{32} + \cdots$$
$$M_{32} = V_{32}(1 + E_2 M_{22}) + V_{33} E_3 M_{32} + \cdots$$
$$M_{42} = \cdots \tag{22}$$

There are some small technical problems in dealing with the disconnected graphs in V_{33} but these can be easily handled. Finally, since the E_j-type Green's functions have positive-definite imaginary parts, all our previous techniques can be applied with almost no change. A detailed discussion will be given later.

We have applied our equations in lowest approximation to the ρ-meson bootstrap problem. We do not find any self-consistent solutions even if selected higher mass states are added to the problem. Further work is in progress on this point.

REFERENCE

1. R. Blankenbecler and R. Sugar, *Phys. Rev.* **136**: B472 (1964).

Equivalent Potential Approach for Strong Interactions

L. A. P. Balázs

TATA INSTITUTE OF FUNDAMENTAL RESEARCH
Bombay, India

In most strong-interaction calculations, the general procedure is to assume some force coming from the exchange of particles and then unitarize it, that is, make the S-matrix unitary. For instance, in N/D calculations,[1] exchange diagrams give an expression for the unphysical singularities; the N/D formulas are then constructed in such a way that they give these singularities correctly and at the same time satisfy unitarity in the physical region.

Actually, the above procedure is the one which is in effect followed in atomic and classical nuclear physics. Here, one takes the Fourier transforms of exchange diagrams to obtain potentials which are then inserted into a Schrödinger equation. Since that equation does give a unitary S-matrix, we have essentially unitarized an exchange. Of course this particular approach is only valid in the nonrelativistic region.

In view of the success of the Schrödinger equation in its domain of validity, it might be desirable to find a relativistic approach which passes over naturally and simply to the ordinary Schrödinger equation (with energy-independent potential) in the nonrelativistic region. That this might be possible within a dispersion approach was first suggested by the work of Charap and Fubini,[2] who constructed a potential by requiring that it reproduce the correct relativistic amplitude at the physical threshold. They showed that their potential was valid over a wide range of energies. They also proposed an explicit iterative scheme for constructing the potential from exchanges.

Our procedure is a simple generalization of the Charap–Fubini approach. We shall require that our potential reproduce the relativistic amplitude at any given energy.† Consider the Schrödinger equation for two particles of equal mass (with mass $= 1$):

$$\nabla^2\psi + [k^2 - V(r, q^2)]\psi = 0 \tag{1}$$

where ψ is the wave function, k is the magnitude of the three-momentum in the center of mass system, r is the radial distance, and $V(r, q^2)$ is the potential which reproduces the relativistic amplitude when $k^2 = q^2$ (we are considering q^2 to be a constant, at least for the time being). We are neglecting spin and exchange forces, which do not change our general arguments. If

$$V(r, q^2) = -\frac{1}{\pi} \int_{t_0}^{\infty} dt' \, v(t', q^2) r^{-1} e^{-r(t')^{1/2}} \tag{2}$$

the physical scattering amplitude $f(k^2, t)$ can be shown to satisfy the Mandelstam representation.[4] This just means that we can write

$$f(k^2, t) = \frac{1}{\pi} \int_{t_0}^{\infty} dt' \frac{f_t(t', k^2)}{t' - t} \tag{3}$$

with

$$f_t(t, k^2) = v(t, q^2) + \frac{1}{\pi} \int_{0}^{\infty} dk'^2 \frac{\alpha(k'^2, t)}{k'^2 - k^2} \tag{4}$$

where f is such that the differential cross section $= |f|^2$, while $t = -2k^2(1 - \cos\theta)$ (θ is the scattering angle in the CM system). From unitarity and equations (3) and (4) it can be shown that[5]

$$\alpha(k^2, t) = \frac{1}{2\pi k} \int_{t_0}^{\infty} dt' \int_{t_0}^{\infty} dt'' \frac{f_t^*(t', k^2) f_t(t'', k^2)}{\mathbf{K}^{1/2}(k^2; t, t', t'')} \theta(t - t_+) \tag{5}$$

where

$$\mathbf{K}(k^2, t, t', t'') = t^2 + t'^2 + t''^2 - 2(tt' + tt'' + t't'') - tt't''k^{-2} \tag{6}$$

and t_+ is the larger zero of $\mathbf{K}$ regarded as a function of t.

Consider the relativistic amplitude $A = (q^2 + 1)f$ when the CM three-momentum is q. Assuming that it satisfies the Mandelstam

---

†A similar approach has been used by Logunov et al.[3] They restricted themselves to perturbation theory, however. They also used an equation which differs from the usual Schrödinger equation with local potential. Most of the considerations given here are also applicable to their equation.

representation,[6] we have

$$A(s, t) = \frac{1}{\pi} \int_{t_0}^{\infty} dt' \frac{A_t(t', s)}{t' - t} \tag{7}$$

where $s = 4(q^2 + 1)$. The lower limit t_0 is the square of the lowest intermediate mass in the crossed t-channel (which is just the process obtained by looking at the scattering diagram sideways). From equations (3) and (7), we see that the requirement that equation (1) give the correct scattering amplitude at $k^2 = q^2$ is equivalent to putting

$$f_t(t, q^2) = 2s^{-1/2} A_t(t, s) \tag{8}$$

Subtracting from equation (4) its value at $k^2 = q^2$, we thus have

$$f_t(t, k^2) = 2s^{-1/2} A_t(t, s) + \frac{k^2 - q^2}{\pi} \int_0^{\infty} dk'^2 \frac{\alpha(k'^2, t)}{(k'^2 - q^2)(k'^2 - k^2)} \tag{9}$$

This, together with equation (5), forms a nonlinear integral equation for $f_t(t, k^2)$, if we are given $A_t(t, s)$. It can be solved by iteration. As a first approximation we drop the integral in equation (9); because of the θ-function in equation (5), this gives f_t correctly for $t_0 < t < 4t_0$. If we now insert this f_t into equation (5), then equation (9) will give an f_t which is exact in the region $t_0 < t < 9t_0$. In general n iterations will give f_t exact for $t_0 < t < (n + 1)^2 t_0$. Once we know f_t and α we can find v from

$$v(t, q^2) = 2s^{-1/2} A_t(t, s) - \frac{1}{\pi} \int_0^{\infty} dk'^2 \frac{\alpha(k'^2, t)}{k'^2 - q^2} \tag{10}$$

which follows from equations (4) and (8).

The above iteration scheme is obviously useful only if we need $v(t, q^2)$ for small t, that is, only if the long-range part of the potential is important. This is, of course, the usual sort of assumption one makes in strong-interaction calculations. In this case, it is not unreasonable if we assume that A_t and α are dominated at large t by a few Regge poles[7] and if the residue and position of these poles are assumed to have only right-hand cuts in the k^2 plane[8]. This is because we can then write

$$f_t(t, k^2) = \frac{1}{\pi} \int_0^{\infty} dk'^2 \frac{\alpha(k'^2, t)}{k'^2 - k^2} \tag{11}$$

which, together with equation (10), leads to $v(t, q^2) = 0$ for large t.[†]

[†]Because of oscillations, only the low k'^2 part of the integral is likely to be important in (11).

Once we know v, we can solve the equation

$$\nabla^2\psi + [q^2 - V(r, q^2)]\psi = 0 \tag{12}$$

to obtain the correct amplitude at the energy corresponding to q^2. Since the above procedure can be carried out for any value of q^2, we can get the amplitude this way at any energy. At $q^2 = 0$, V is the same as the potential obtained by Charap and Fubini.[2] Since they showed that $V(r, 0)$ gives good results over a wide range of energies in the nonrelativistic region, we see that $V(r, q^2)$ must be slowly varying in that range.

To implement the above procedure, we must have a way of calculating the absorptive part A_t. For this purpose we shall use a simplified version of the strip approximation.[9] We first approximate A_t by keeping several partial waves in the t-channel:

$$A_t^{(0)}(t, s) = \sum_{l=0}^{L} (2l + 1) \operatorname{Im} A_l(t) P_l\left(1 + \frac{2s}{t - 4}\right) \tag{13}$$

where A_l is a partial-wave amplitude. Since A has been assumed to satisfy the Mandelstam representation

$$A_t(t, s) = A_t^{(0)}(t, s) + \frac{1}{\pi} \int_4^\infty ds' \rho(s', t)$$

$$\times \left[\frac{1}{s' - s} - \frac{2}{t - 4} \sum_{l=0}^{L} (2l + 1) Q_l\left(1 + \frac{2s'}{t - 4}\right) P_l\left(1 + \frac{2s}{t - 4}\right)\right] \tag{14}$$

where we have subtracted out the lowest-order term. This has the effect of suppressing the contribution of the third double-spectral function which would otherwise have to be included in equation (14). It should also suppress the high s' part of the integral in equation (14) and so we need consider only elastic s-channel unitarity. This when combined with equations (7) and (14) gives[5]

$$\rho(s, t) = \frac{1}{\pi q s^{1/2}} \int_{t_0}^\infty dt' \int_0^\infty dt'' \frac{A_t^*(t', s) A_t(t'', s)}{K^{1/2}(q^2; t, t', t'')} \theta(t - t_+) \tag{15}$$

for the double spectral function ρ.

Equations (14) and (15) closely resemble (4) and (5), and can be solved in exactly the same way. In lowest order, we drop the integral term in (14). Then, because of the θ-function in equation (15), we will have A_t in the region $t_0 < t < 4t_0$. If we insert this first approximation into equation (15), then (14) will give A_t in the region $t_0 < t < 9t_0$; in general, equation (14) will diverge, so we must either put in a cut-off or subtract out additional partial waves. We can repeat this procedure any number of times to obtain A_l for larger and larger values of t. Once we have A_l, we also have v. If we go to the same order in the iterations of both equations (4) and (5) and (14) and (15), we moreover obtain real v. This can be seen from equations (5) and (15), which, together with (8), give

$$2s^{-1/2}\rho(s, t) = \alpha(q^2, t) \tag{16}$$

Equations (10) and (14) then give Im $v(t, q^2) = 0$.

The above scheme, of course, presupposes a knowledge of the low partial waves in equation (13). These, however, can be found by solving (12) in the crossed channel. Because of the subtractions in equation (14), we would expect the integral term in (14) to be suppressed. On the other hand, there is no corresponding suppression of the integral in (10). Since the latter corresponds to a repulsive contribution to v, at least in the lowest nonzero approximation, this suggests that there may be a repulsive potential at moderately small distances. Such a potential would further suppress any large t effects. It can be shown, however, that although it may be important for small s this contribution becomes small as we go to higher energies.

So far, no calculations have been made keeping the integral terms in equations (9), (10), and (14). A very crude calculation has, however, been made of the ρ-meson in $\pi\pi$-scattering in which only the ρ contribution to (13) was kept. This gives in the $I = 1$ state

$$A_l(t, s) = 3\beta_{11} \text{ Im } A_1(t)p_1\left(1 + \frac{2s}{t - 4}\right) \tag{17}$$

where $\beta_{11} = \frac{1}{2}$ is the crossing matrix element connecting the $I = 1$ state in the t-channel to the $I = 1$ state in the s-channel. We shall, as usual, make a delta-function approximation for the resonance

$$\text{Im } A_1(t) = 4\pi\Gamma_1 q_R^2\delta(t - m^2) \tag{18}$$

such that the integral over equation (18) is the same as that over the resonance. Here $(2q_R^3\Gamma_1/m)$ is the half-width in the q^2 variable, $q_R^2 = \frac{1}{4}m^2 - 1$, and m is the mass of the ρ. Combining equations (2), (10), (17), and (18), we obtain in lowest order

$$V(r, q^2) = -24\beta_{11}\Gamma_1 s^{-1/2}(s + 2q_R^2)r^{-1}e^{-mr} \tag{19}$$

with the extra factor of two coming from the fact we have contributions from both t- and u-channels.

To solve equation (12), we use the standard expression[10] for the phase shift δ_1

$$\cot \delta_1 = \left[q \int_0^\infty dr\, V u_1^2 - 2 \int_0^\infty dr\, qr\, n_1(qr) V(r, q^2) u_1(r) \right.$$

$$\left. \times \int_0^r dr'\, qr'\, j_1(qr') V(r', q^2) u_1(r') \right] \left[\int_0^\infty dr\, qr\, j_1(qr) V u_1 \right]^{-2} \quad (20)$$

which is stable with respect to variations in the wave function u_1 and independent of its normalization. We shall thus take the crude form

$$u_1(r) = r^2 e^{-mr} \quad (21)$$

which has the correct behavior at small r and roughly represents the trapping of the particle within the range of the force, a situation we would expect for a resonance. Finally we make an effective range expansion about $q^2 = 0$ to obtain

$$2s^{-1/2} q^3 \cot \delta_1 = a_e - r_e q^2 \quad (22)$$

where

$$a_e = m^3 \left[\frac{128m}{81 \Gamma_1(m^2 + 4)} - \frac{1}{2} \right] \quad (23)$$

and

$$r_e = \frac{m}{32}(37 - 8m^2) - \frac{128m^2}{81 \Gamma_1(m^2 + 4)} \left[1 - \frac{8m^2}{m^2 + 4} \right] \quad (24)$$

Equation (22) will give a resonance at $q^2 = a_e r_e^{-1}$ with a reduced width $\Gamma_1 = r_e^{-1}$. The latter relation can be trivially solved to give Γ_1 in terms of m, as can be seen from (24). Using this result we can then try various values of m and see whether the condition $q_R^2 = a_e r_e^{-1}$ is satisfied. Approximate self-consistency is achieved with $m = 3.9$ and $\Gamma_1 = 0.55$, which correspond to a mass of 545 MeV and a width of 160 MeV. The experimental values are $m = 5.5$ and $\Gamma_1 = 0.16$ which correspond to a mass of 765 MeV and a width of 100 MeV.

The above results may, of course, be misleading because of the drastic approximation that had to be made to make a hand-calculation possible. More careful calculations are now in progress.

REFERENCES

1. G. F. Chew and S. Mandelstam, *Phys. Rev.* **119**: 467 (1960).
2. J. M. Charap and S. P. Fubini, *Nuovo Cimento* **14**: 540 (1959); **15**: 73 (1960).
3. A. A. Logunov and A. N. Tavkhelidze, *Nuovo Cimento* **29**: 380 (1963); A. A. Logunov, A. N. Tavkhelidze, I. T. Todorov, and O. A. Khrustalev, *Nuovo Cimento* **30**: 134 (1963).
4. R. Blankenbecler, M. L. Goldberger, N. N. Khuri, and S. B. Treiman, *Ann. Phys.* **10**: 62 (1960).
5. See, for instance, G. F. Chew, *S-Matrix Theory of Strong Interactions*, W. A. Benjamin, Inc., New York, 1961, Chapter 7.
6. S. Mandelstam, *Phys. Rev.* **112**: 1344 (1958).
7. T. Regge, *Nuovo Cimento* **14**: 951 (1959).
8. V. Singh, *Phys. Rev.* **127**: 632 (1962).
9. G. F. Chew and S. C. Frautschi, *Phys. Rev.* **123**: 1478 (1961).
10. See, for instance, P. M. Morse and H. Feshbach, *Methods of Theoretical Physics, Part II*, McGraw-Hill Book Co., Inc., New York, 1953, p. 1128.

"Repulsive" Potential Approach to Pion Resonances

A. N. MITRA

UNIVERSITY OF DELHI
Delhi, India

I shall be concerned here with a phenomenological approach to pionic interactions that we have been pursuing during the last year in order to understand some of the pion resonances. This approach, which is characterized by low-energy repulsion and high-energy attraction, was motivated through certain manifestations of s- and p-wave π-π interactions. For s-waves, some important manifestations are pion–nucleon scattering[1] and τ-decay anomalies. (The results of various approaches to this problem have been summarized by Kacser.[2]) For p-waves resonances such as ρ, ω, and ϕ come first to mind.

1. s-WAVE INTERACTION

For s-wave interactions, the so-called ABC particle, which is merely a nonresonant $T = 0$ interaction with a large scattering length, provides an explanation of s-wave π-N scattering,[1] but does not explain the anomalies in τ-decay.[2] On the other hand, the (fictitious) σ-particle, which is a $T = 0$, π-π resonance at about 400 MeV with a large width of approximately 100 MeV,[3] seems to play a useful role in the understanding of τ-decay,[4] K_1^0-K_2^0 mass difference nucleon–nucleon phase shifts,[6] etc. However, experimental evidence in favor of such a particle seems to have been rather shaky,[7] since it was first postulated.[8] It is possible that such a large width of the particle, disproportionate to its mass, makes its experimental detection extremely difficult. It may, however, be interesting to see what kind of mechanism, if any, can give rise to such a particle.[9] (Brown has discussed this question within the frame work of dispersion theory.)

A model of $T = 0$, π-π resonance similar to the σ-particle was suggested by the author,[10] in the context of $K\pi3$-decay, at the Rochester Conference 1960. However, the dispersion theoretic results on τ-decay at that time,[11] which indicated a repulsive force in $T = 0$, seemed to contradict this conclusion. Presumably, for this reason this "resonance model" did not attract much attention until the model of Brown and Singer[3] appeared. More recently, a closer examination of this "old" resonance model showed that it could provide as satisfactory an explanation of τ- and η-decay anomalies[12] as the pion-pole model of p-wave resonances.[2] A possible way of distinguishing between the predictions of the resonance and pion-pole models is to look at the angular distribution of the unlike pion, for which a nearly isotropic distribution is predicted by the resonance model and an anisotropic one by the pion-pole model.

From the mathematical point of view, a possible way to reconcile the dispersion theoretic result of a repulsive $T = 0$ interaction[11] with a π-π resonance[3,10] in $T = 0$ would be to invoke the dominance of the dipion pole in a dispersion formula, together with a constructive interference with background integral.[12] Some time ago we looked at this question within the framework of a Schrödinger equation for the final 3π-state in τ-decay.[13] We found that a $T = 0$ interaction, taken as *repulsive* at low energies, could explain the high-momentum bias of the unlike pion. This result is, in principle, consistent with the above statement (of pole-dominance in a dispersion formula with constructive interference with the background integral) provided that such a repulsive interaction can at the same time give rise to a π-π resonance. It is clear, of course, that a purely attractive interaction which has no centrifugal barrier in the s-state cannot give rise to a resonance, so that an ABC anomaly cannot be the answer to a σ particle. On the other hand, an interaction which is repulsive at low energies, may or may not have an inner attractive well. If the potential is *local*, it is of course possible to recognize these features purely from inspection. However, with a nonlocal potential, one must either look at the detailed behavior of the wave function or examine the energy dependence of the phase shifts to infer its detailed spatial structure.

Now, the potential considered in reference 12 was of the form

$$\langle \mathbf{p}|V|\mathbf{p}'\rangle = \lambda g(p)g(p') \qquad (1)$$

$$g(p) = (\beta^2 + p^2)^{-1} \qquad (2)$$

where $\lambda > 0$ corresponds to repulsive interaction. A detailed examination of the phase shifts from such an interaction,[14] using a Schrödinger-type equation

$$(\omega_p^2 - \omega_K^2)\psi(\mathbf{p}) = -\int\langle\mathbf{p}|V|\mathbf{p}'\rangle\psi(\mathbf{p}')d\mathbf{p}' \qquad (3)$$

where $2\omega_K$ is the total CM energy, reveals indeed that it is capable of generating a reasonably high energy resonance with a width *decreasing* with the range β^{-1} of the potential. This property is exactly the opposite of what is exhibited by a purely attractive interaction, for which the width increases with the range. Thus, in this model, the mechanism of resonance formation depends more on the effectiveness of the barrier than on the strength of the inner attraction (which of course must be present if a resonance has to develop at all. If such a picture is taken seriously, it is clearly possible to reconcile the earlier conflicting results on the effect of final state π-π interaction on $K3\pi$- or η-decays. The model is, of course, highly qualitative, yet its physical consequence looks sufficiently interesting to warrant further examination. So we now turn to the corresponding p-wave model.

2. p-WAVE INTERACTION (2π-STATES)

The most obvious manifestation of a p-wave π-π interaction is perhaps the ρ-meson. However, its inordinately large mass for a mere p-wave resonance is in striking contrast to the more familiar 3π-resonance at 1238 MeV. The usual answer to this anomaly in the language of dispersion theory is, of course, that the ρ-meson is a multichannel problem, so that channels other than the π-π one have a nontrivial effect on its energy and width. For our present interest, the question is whether it is possible to give some effective representation to the p-wave interaction so as to incorporate the physical effects of several channels, and, if so, what its qualitative features should be. Our experience with the corresponding s-wave case strongly suggests that a high mass and a moderate width of the ρ-meson go rather well with a potential which is repulsive (*in addition* to the centrifugal barrier) at longer distances but attractive at shorter ones. With a purely attractive interaction, on the other hand, one would need an extremely short-ranged force to hold such a high-energy resonance ($m_\rho \sim 760$ MeV) for a reasonable amount of time ($\Gamma_\rho \sim 100$ MeV).

We are thus led to consider a p-wave π-π potential of the form[15]

$$\langle \mathbf{p}|V|\mathbf{q}\rangle = \lambda v(p)v(q)(\mathbf{p}\cdot\mathbf{q}) \qquad (\lambda > 0) \tag{4}$$

$$v(p) = \exp\left(-\frac{p^2}{2\beta^2}\right) \tag{5}$$

through a semirelativistic, Schiff-type[16] Schrödinger equation of the form

$$2(p^2 - K^2)\psi(\mathbf{p}) = -\lambda \int d\mathbf{q}\, v(p)v(q)\mathbf{p}\cdot\mathbf{q}\,\psi(\mathbf{q}) \tag{6}$$

$$4K^2 = E^2 - 4\mu^2 \tag{7}$$

where E is the total CM energy. This leads to the resonance condition[15]

$$1 + 2\pi\lambda P \int_0^\infty dq\, q^4 \exp\left(\frac{-q^2}{\beta^2}\right)(q^2 - K_\rho^2)^{-1} = 0 \tag{8}$$

$$K_\rho^2 = \tfrac{1}{4}(m_\rho^2 - 4\mu^2) \approx 6\mu^2 \tag{9}$$

and the full width (at resonance)

$$\Gamma_\rho \approx 2\pi^2\lambda\omega_{k_\rho}^{-1}K_\rho^3 v^2(K_\rho)(K_\rho^2 - \tfrac{5}{2}\beta^2) \tag{10}$$

which works out as 130 MeV and 69 MeV for $\beta = 1.0\,\mu$ and 0.9 μ, respectively.

It may be noted that if an attractive interaction ($\lambda < 0$) were used in (4) to (6), one would need a highly artificial value of β, namely, $\beta \sim 30\mu$, in order to get $\Gamma_\rho \sim 100$ MeV at the observed mass of the ρ-meson. Thus, it seems that an apparently repulsive interaction like (6) gives a much easier fit to the ρ-meson parameters than a purely attractive force. We would like to interpret this result to imply that perhaps multichannel effects are more satisfactorily represented by (4) with $\lambda > 0$ than with $\lambda < 0$.

This model has a simple SU_3 ramification,[17] as well. Thus, if the interaction (4) is phenomenologically extended to K-π interactions, using approximate SU_3 symmetry, then, within the octet version, $\lambda_{K\pi} = \tfrac{3}{4}\lambda_{\pi\pi}$, where $\lambda_{\pi\pi}$ is the coupling strength appearing in (4). If we now use a Schrödinger equation such as (6) for the K-π interaction, where the relative momentum K is related to the total energy E of the π-K system by

$$(K^2 + m_\pi^2)^{1/2} + (K^2 + m_K^2)^{1/2} = E \tag{11}$$

then it is clear from (8) that the effective value of the resonance momentum $K_{K^*}^2$ of the K^*-resonance (888 MeV) would be lower than K_ρ^2 if $\lambda > 0$, higher if $\lambda < 0$. Experimentally, at $E = m_{K^*} = 888$ MeV, $K_{K^*}^2 \approx 4.34\, \mu^2$. The predicted value with $\lambda > 0$ is $K_{K^*}^2 = 4.78\, \mu^2$, which is at least in the right direction. With $\lambda < 0$, on the other hand, $K_{K^*}^2 \approx 7.5\, \mu^2$, which is in total discord with observation. Thus, it appears that our model gives qualitatively correct results even when it is extended to kaon interactions through SU_3 symmetry.

We shall next consider some 3π- and 4π-states on this model.

3. p-WAVE INTERACTION (3π- AND 4π-STATES)

As examples of three-pion systems, the interesting objects of study are the ϕ- and ω-mesons and the more recently discovered A_1- and A_2-mesons.[18] An important difference between ϕ and ω, both of which have the same quantum numbers $J^P T^G = 1^- 0^-$, is the lack of 3π-modes of the former, despite the advantages of phase space. Even the decay $\phi \to \pi + \rho$ is rare compared with $\phi \to K + \bar{K}$, again in spite of phase space. It may be interesting to ask whether our model can accommodate these features in ϕ without, for instance, invoking new selection rules such as the A-parity of Broznan and Low.[19]

The formulation of the 3π problem using separable potentials, but with relativistic kinematics, was given sometime ago[20] by the author. We have used this in the Schiff approximation[16] to study the effect of an interaction like (4) (with $\lambda > 0$), on a 3π-system[15] with $J^P T^G = 1^- 0^-$. The Schrödinger equation

$$D(E)\Psi = -(V_{12} + V_{23} + V_{31})\Psi \tag{12}$$

when

$$D(E) = P_1^2 + P_2^2 + P_3^2 - 2K_0^2 \tag{13}$$

and

$$6K_0^2 = E_0^2 - 9\mu^2 \tag{14}$$

where E_0 is the total 3π energy leads, for the $1^- 0^-$ state, to the wave function

$$\Psi = D^{-1}(E) \sum_{ijk} (\mathbf{p}_{ij} \times \mathbf{P}_k) v(p_{ij}) F(P_k) \tag{15}$$

where

$$2\mathbf{p}_{ij} = \mathbf{P}_i - \mathbf{P}_j, \qquad \mathbf{P}_i + \mathbf{P}_j + \mathbf{P}_k = 0 \tag{16}$$

$F(P)$ is the π-p wave function satisfying the approximate equation[15]

$$(P^2 - K_1^2)F(p) \approx -\lambda' \int_0^\infty q^4 \, dq \, F(q) u(p) u(q) \tag{17}$$

$$\tfrac{3}{4}K_1^2 = K_0^2 - K_\rho^2 \tag{18}$$

$$u(p) = \exp\left(-\tfrac{5}{8}\frac{p^2}{\beta^2}\right) \qquad \lambda' = \tfrac{16}{9}\beta^{-5}\pi^{-3/2}(K_\rho^2 - \tfrac{5}{2}\beta^2) \tag{19}$$

which has the form of a scattering equation for $K_0^2 > K_\rho^2$, that is, $E_0 > m_\rho + m_\pi$. Equation (17) is now analogous to (6), and one finds the resonance energy E_R and the width Γ_R as

$$(E_R, \Gamma_R) \approx (1230, 12.5) \text{ MeV and } (1240, 5.2) \text{ MeV} \tag{20}$$

for $\beta = 1.0 \; \mu$ and $0.9 \; \mu$, respectively. It would be natural to interpret this resonance in terms of ϕ (1020), since this particle, rather than ω, is a resonant state of ρ and π. Of course, the calculated mass is too high for the observed one, but it is a nontrivial result that the width is of the right order of magnitude for ϕ, without any extra assumption beyond "repulsive" π-π interaction of the type (4).[†]

Such a mechanism cannot, of course, be the right candidate for ω, since a bound state of π and ρ cannot be brought about by this model. Thus, ω is in principle excluded from our scheme. This may not be too undesirable in the context of certain conjectures[21] that ϕ, rather than ω, might belong to the vector octet, with little mixing. Such an assignment also helps in keeping alive Sakurai's idea of universal coupling of ω to the baryonic charge.[23]

To speculate further with this model, the replacement of a pion by a kaon in the 3π-system leads to a $K\pi\pi$ resonance of $J^P = 1^-$ and $T = \tfrac{1}{2}$, as the analog of ϕ. A very crude calculation indicates a mass of 1300 MeV, using the SU_3 versions of the effective π-π coupling constants. Though a number of $K\pi\pi$ resonances have now been reported,[23] we are rather inclined to associate this resonance with the one at 1175 MeV,[25] and it would be interesting to know its detailed quantum numbers.

[†]It may be tempting to conclude that the small width is due to the p-wave centrifugal barrier. We wish to emphasize, however, that the mechanism responsible for bringing this about is not so much the centrifugal barrier as the main π-π interaction itself, to which the former is merely a correction.

Finally, we have looked into the energy of a 4π-state of $T = 0$ and $J^P=0^-$, 1^+ using the interaction (4).[25] Our result shows that a resonance or bound state in the 1^+ state is not possible on this model. Also ruled out is a bound 4π-state of 0^-, so that the model does not predict the η-meson. However, the model predicts a 4π-resonance of $J^P T^G = 0^- 0^+$ at an energy very close to that of the χ^0-particle,[26] whose quantum numbers are most likely to be $0^- 0^+$. A second resonance at approximately 1400 MeV with $J^P T^G = 0^- 0^+$ is also predicted by the model, though its precise interpretation in terms of physical objects is not clear to us.[26]

4. CONCLUSION

We have considered some physical consequences of the assumption that the π-π force has a long-range repulsion and a short-range attraction. The assumption can give rise to an s-wave resonance such as the σ-particle, and as such helps in the understanding of $K\pi3$- and η-decays, K_1^0-K^2 mass difference, etc. It is capable of generating a high-energy p-wave resonance with a moderate width, the magnitude of the resonance momentum *increasing* with the strength of the interaction. Thus, it provides a qualitative understanding of the positions of the ρ- and K^*-mesons through approximate SU_3 symmetry. It also gives a qualitatively satisfactory mechanism for the ϕ-meson and at least one of the recently discovered $K\pi\pi$ resonances. Finally, it generates a 4π-state of $0^- 0^+$ at an energy appropriate to the χ^0-meson.† The mathematical model used in making these inferences is, of course, extremely crude and open to several objections such as inadequate treatment of relativistic effects, etc. However, the qualitative results appear sufficiently interesting to warrant more accurate calculations along similar lines.

REFERENCES

1. J. Hamilton, in CERN Conference on High Energy Physics (1962).
2. C. Kacser, *Phys. Rev.* **130**: 335 (1963).
3. L. M. Brown and P. Singer, *Phys. Rev. Letters* **8**: 460 (1962).
4. L. M. Brown and P. Singer, *Phys. Rev.* **133** B: 3812 (1964).

---

†This is not to suggest that the principal mode of decay of χ^0 should be 4π rather than the observed $\eta + 2\pi$.

5. K. Nishijima, *Phys. Rev. Letters* **12**: 39 (1964).
6. Scotti and D. Y. Wong, *Phys. Rev. Letters.*
7. Y. Nikitin, in Proceedings of the Dubna Conference on High Energy Physics (1964).
8. N. Samios *et al.*, *Phys. Rev. Letters* **9**: 139 (1962).
9. L. M. Brown, *Phys. Rev. Letters* (1964).
10. A. N. Mitra, in Proceedings of the Rochester Conference on High Energy Physics (1960).
11. N. Khuri, Proceedings of the Rochester Conference on High Energy Physics (1960).
12. A. N. Mitra and S. Ray, *Phys. Rev.* **135** B: 146 (1964).
13. A. N. Mitra and S. Ray, *Ann. Phys.* (N.Y.) **21**: 439 (1963).
14. A. N. Mitra, *Nuovo Cimento* **32**: 506 (1964).
15. A. N. Mitra, *Nuovo Cimento* **33**: 1220 (1964).
16. L. I. Schiff, *Phys. Rev.* **125**: 777 (1962).
17. A. N. Mitra, *Phys. Letters* **12**: 61 (1964).
18. R. L. Lander *et al.*, *Phys. Rev. Letters* **13**: 346 (1964).
19. W. Broznan and F. E. Low, *Phys. Rev. Letters* **12**: (1964).
20. A. N. Mitra, *Phys. Rev.* **127**: 1342 (1962).
21. Pignotti *et al.*, *Physics Letters* **9**: 273 (1964).
22. J. J. Sakurai, *Ann. Phys.* **11**: 1 (1960).
23. R. Armenteros, in Proceedings of the Dubna Conference on High Energy Physics (1964).
24. T. Wangler *et al.*, *Phys. Letters* **9**: 71 (1964).
25. A. N. Mitra and S. Ray, *Phys. Rev.* **137**: 4B (1965)
26. G. R. Kalbfleisch *et al.*, *Phys. Rev. Letters* **12**: 527 (1964).

A Model of a Unitary S-Matrix for Peripheral Interactions

K. DIETZ

THEORETICAL PHYSICS DIVISION, CERN
Geneva, Switzerland

In this discussion we are going to deal with particle scattering at a few GeV, say, 2 to 5 GeV. At these energies, we certainly will have to deal with a many-channel problem; however, one of the prominent features found in experiments is the abundance ($\sim 50\%$) of quasi-two-body final states: for example,

$$K^+ p \rightarrow K^* N$$
$$KN^*$$
$$K^* N^*$$

An immediate interest in these collisions arises from the fact that these reactions are used to establish baryon and meson spectra. Furthermore, these reactions can possibly be used to extract information on the couplings of these particles or resonances to each other; that is, on the form of the effective couplings and the values of the corresponding coupling constants.

Another important feature of these quasi-two-body reactions is that they all work best with small momentum transfer: The differential cross section is a rapidly decreasing function of the momentum transfer (Fig. 1). This immediately tells us that many-particle waves come into play (around fifty for 3 GeV/c $K^+ p$ reactions).

Nonrelativistic quantum mechanics may serve us as a guide for a possible treatment of the problem. With it, one can prove[1] that high partial waves $t_J(s)$ for sufficiently high energies are given by

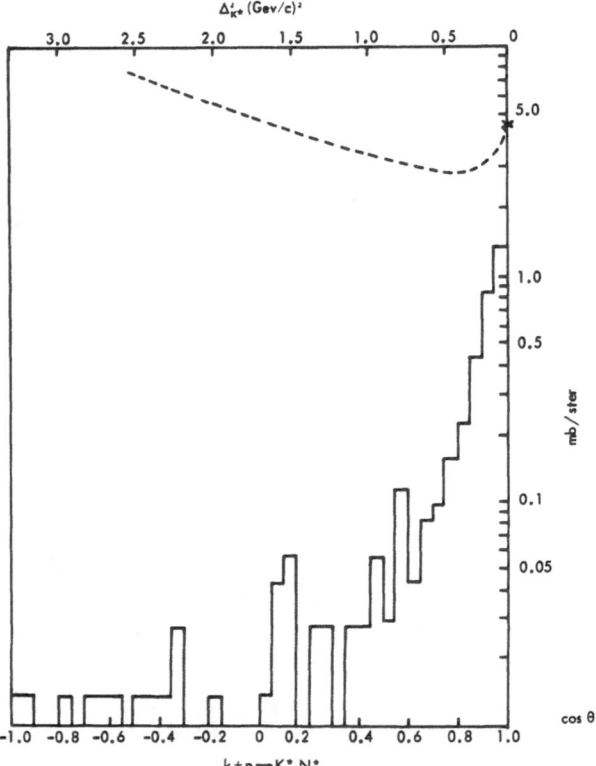

Fig. 1. Differential cross section for the reaction $K^*p \rightarrow K^*N^*$ at 3 GeV/c. (Taken from Ferro-Lussi *et al.* "The reaction at 3 GeV/c and the production mechanism of k^*N^*," CERN preprint.) ... cross section calculated from the unadorned π-exchange graph.

the Born approximation

$$t_J(s) \rightarrow t_J^{\text{Born}}(s) \tag{1}$$

for $J \rightarrow \infty$ and $s > s_0$, fixed.

This relation holds true even for potentials which are singular like $1/r^N$ at the origin.[2] Singularities of that type are found in field-theoretic potentials if the particles involved have higher spins.

Our assumption will be that this relation holds true in a theory capable of describing the phenomenon we are looking at. One might call this a *peripheral* model, because in an impact parameter language this assumption evidently means that the forces at long relative distances of the scattered particles are given by the Born approximation. More specifically, that means that the high partial waves for a scattering process

$$a + b \rightarrow c + d$$

can be obtained from a graph

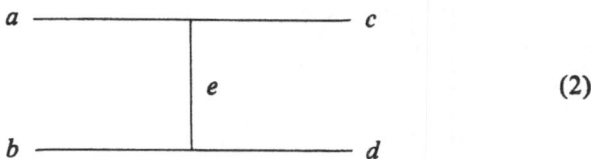

(2)

If we accept that, then the physical question mentioned before is just to find out from experiment what kind of particle e is exchanged and how it is coupled.

Now, that would all be fine if it were possible to determine high partial waves from experiment, but, needless to say, this is very far from being achieved. The only quantities available for the analysis are differential cross sections, polarizations, and, of course, polarization correlations. To predict these, we certainly need to know all partial waves. So the main and in fact the really difficult problem is to calculate the low partial waves. The first idea, to obtain the latter also from a one-particle exchange (OPE) graph, does not work. The reason can be seen if we plot (Fig. 2) the following function:

$$C_J = K(s) \sum_{\lambda_a \lambda_b \lambda_c \lambda_d} |\langle \lambda_c \lambda_d | T^J_{(s)} | \lambda_a \lambda_b \rangle|^2 \tag{3}$$

where λ_a, λ_b, λ_c, and λ_d are the helicities of particles a, b, c, and d; $\langle \lambda_c \lambda_d | T^J_{(s)} | \lambda_a \lambda_b \rangle$ is the helicity amplitude for the process $a + b \rightarrow c + d$; and $K(s)$ is a kinematical factor such that unitarity restricts $|C_J| \leqslant 1$ for fixed s as a function of J.

We see that the low partial waves calculated from the Born approximation violate unitarity in a striking manner. This well-known difficulty is the main reason that the differential cross section calculated from (2) usually turns out to be much too broad (see Fig. 1). The first remedy, proposed by Sopkovitch[3] and later extended to include spin by Jackson and Gottfried[4] and Chiu and Durand,[5] was to damp the low Born partial waves by including the distortion of the wave functions in the initial and final states. This distortion was thought to be caused by an optical potential resulting from the presence of many open channels. Under the assumption that the scattering in initial and final states is described by one and the

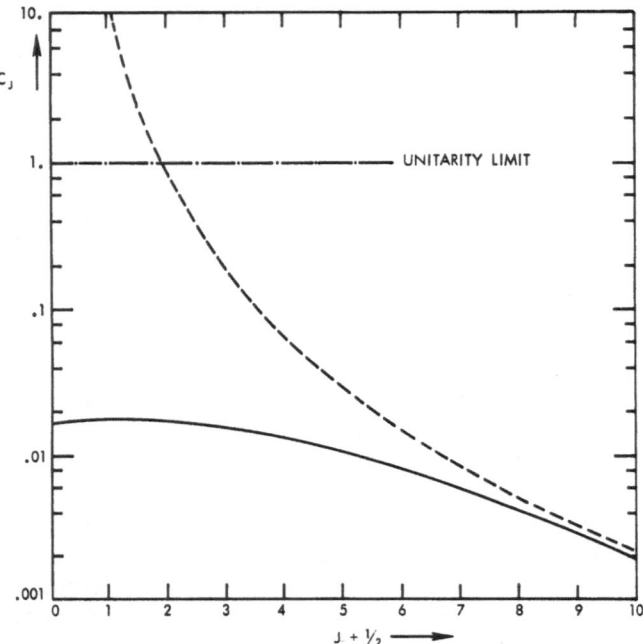

Fig. 2. C_J as a function of $J + \frac{1}{2}$ for the reaction $\pi^- p \to \rho^- p$ at 4 GeV/c. (Taken from K. Gottfried and J. D. Jackson, CERN preprint TH. 428.)

same purely imaginary potential, one derives in the high-energy (WKB) approximation the expression for distorted Born matrix element:

$$\langle \lambda_c \lambda_d | \underset{ab \to cd}{T^J}(s) | \lambda_a \lambda_b \rangle = e^{-2\chi_J} \langle \lambda_c \lambda_d | T^J_{\text{Born}}(s) | \lambda_a \lambda_b \rangle \tag{4}$$

The phase shift $i\chi_J$ describes the scattering in the initial and final states and is purely imaginary. Another important point is that the elastic scattering amplitude is assumed to be spin independent and purely imaginary. New experiments[6] have indicated that the latter assumption is not true. The first point waits for experimental clarification.

    Now if the reaction $a + b \to c + d$ is only a small fraction of the total cross section, χ_J can be determined from, say, the elastic ab scattering and is therefore an experimentally given quantity

$$e^{-2\chi(\rho)} = 1 - \frac{\sigma_T}{4\pi b} e^{-\rho^2/2b} \tag{5}$$

where $\rho = J/k$ is the impact parameter, k is the center-of-mass system momentum, and σ_T and b are experimentally given.

Formula (4) now contains no additional free parameters. The calculation showed good results for the reaction $\pi^- p \to \rho^- p$ λ and $K^+ p \to K^* p$ for both differential cross sections and the decay correlations, that is, the polarization of the produced unstable particles. It worked rather badly for, say, $K^+ p \to K^* N^*$ and $N\bar{N} \to N^* \bar{N}^*$. In these reactions, a violent violation of unitarity occurs, and the damping due to the distortion factor does not seem to be sufficient. In the attempt to get a stronger damping, a different scattering in the initial and final states was assumed; however, it was not very satisfactory, and also the derivation of the generalized formula

$$\langle \lambda_c \lambda_d | T^J(s) | \lambda_a \lambda_b \rangle = e^{-[\chi_J \,(\text{initial}) + \chi_J \,(\text{final})]} \langle \lambda_c \lambda_d | T^J_{\text{Born}}(s) | \lambda_a \lambda_b \rangle \qquad (6)$$

is possible only if the direct potential (that is, the Born potential) is assumed to have a much smaller width than the optical potential. Therefore, the model with different initial and final state distortion represented by (6) does not seem to be applicable to the physical situation.

Pilkuhn and myself[7] have been trying to find a procedure to "unitarize" the Born terms (2). It is clear that such a procedure should include all present channels, i.e., all quasi-two-body reactions and, because these make up only 50% of the total cross section, also three- and more-body reactions.

Our considerations will make extensive use of the K-matrix. We therefore have to discuss first the parameterization of the S-matrix in terms of the K-matrix. The dynamics of the model will be described in a second section which displays the assumptions on the K-matrix elements and discusses the model.

We introduce the K-matrix by

$$S = \frac{1 + iK}{1 - iK} \qquad (7)$$

where S is unitary if K is Hermitian. Furthermore, if time-reversal invariance holds, phases can be chosen such that S is symmetric. K is then real and symmetric and has as many matrix elements as the number of independent parameters necessary to write down the most general unitary S-matrix. Therefore, the K-matrix is an efficient means for the parameterization of the S-matrix. From (7) we immediately get Heitler's integral equation for $T = -i(S - 1)$:

$$T = 2K + iKT \qquad (8)$$

which we are now going to solve for T assuming the presence of two- and three-body channels only. This equation is much simpler to handle if we introduce partial waves.

(I) For a two-body reaction $a + b \rightarrow c + d$ we introduce the matrix T_{22} labeled by the helicities $\lambda_a, \lambda_b, \lambda_c$, and λ_d of the particles a, b, c, and d, and, if several two-body channels are present, by the channels indices i, j:

$$T_{22} = (\langle \lambda_c \lambda_d | T^J_{(s)} | \lambda_a \lambda_b \rangle)_{ik} \qquad (9)$$

(II) For $2 \rightarrow 3$ particle process $a + b \rightarrow c + c' + d$, we introduce the matrices $T_{23}(\sigma)^8$ labeled by helicities, by total angular momentum l, helicity Λ, and invariant mass σ of a two-particle (in our case 2-meson) subsystem of the three-particle state, channel indices i, k, etc.

$$T_{23} = (\langle \lambda_c \lambda_{c'} \lambda_d | T^{J;l;\Lambda}_{(s,\sigma)} | \lambda_a \lambda_b \rangle)_{ik} \qquad (10)$$

(III) Similarly we introduce the matrix $T_{32}(\sigma)$ for a $3 \rightarrow 3$ particles process and $T_{33}(\sigma, \sigma')$ for the process $a + a' + b \rightarrow c + c' + d$

$$T_{33}(\sigma, \sigma') = (\langle \lambda_c \lambda_{c'} \lambda_d | T^{J;l,\Lambda;l',\Lambda'}_{(s,\sigma';\sigma)} | \lambda_a, \lambda_{a'}, \lambda_b \rangle)_{ik}$$

Writing equation (8) in terms of these helicity amplitudes, we find that

$$T_{22} = 2K_{22} + i\rho_2 K_{22} T_{22} + i \int d\sigma' \rho_3(\sigma') K_{23}(\sigma') T_{32}(\sigma')$$

$$T_{32}(\sigma) = 2K_{32}(\sigma) + i\rho_2 K_{32}(\sigma) T_{22} + i \int d\sigma' \rho_3(\sigma') K_{33}(\sigma', \sigma) T_{32}(\sigma')$$

where ρ_2 and $\rho_3(\sigma)$ are the two- and three-particle phase space factors. Unitarity and symmetry of the S-matrix are guaranteed if the K's are real, and

$$K_{22} = K^T_{22}$$
$$K_{23}(\sigma) = K^T_{32}(\sigma) \qquad (13)$$
$$K_{33}(\sigma, \sigma') = K^T_{33}(\sigma, \sigma')$$

where K^T is the transposed matrix. Defining a matrix product by

$$\int d\sigma' \rho_3(\sigma') K_{23}(\sigma') T_{32}(\sigma') \equiv K_{23} \times T_{32} \qquad (14)$$

we can write down a solution for the submatrix T_{22} which we are interested in:

$$T_{22} = 2(K_{22} + B_{22})[1 - i\rho_2(K_{22} + B_{22})]^{-1} \tag{15}$$

where

$$B_{22} = iK_{23} \times (1 - iK_{33})^{-1} \times K_{32} \tag{16}$$

It might be worthwhile to check unitarity in the case where only one two-particle channel is present. Then unitarity means that

$$S_{22} = 1 + i\rho_2 T_{22} = \frac{1 + i\rho_{22}(K_{22} + B_{22})}{1 - i\rho_{22}(K_{22} + B_{22})} \tag{17}$$

has a modulus less than or equal to one. From (17) it follows immediately that $|S_{22}| \leqslant 1$ if and only if Im $B_{22} \geqslant 0$. Looking at (16) one sees that

$$\text{Im } B_{22} = K_{23} \times (1 + (K_{33})^2)^{-1} \times K_{32} \tag{18}$$

is non-negative because the symmetry conditions (13) tell us that K_{33}^2 is a positive operator and that therefore Im B_{22} is non-negative.

So for our formal considerations, we are now going to describe how the solution (15) can be used for setting up a model for our problem.

As two-body channels we consider, besides the real two-particles states, also the quasi-two-body state, *i.e.*, resonance + particle or resonance + resonance state. This means, that in (12) we perform the integration over the resonance contributions to the integrals which are expected to have a sharply peaked σ-variation, and count them as two-particle channels. The remaining nonresonant contributions to the integrands are expected to be smoothly varying functions of σ. The validity of such an approach is always assumed when experimentalists try to separate effective mass distributions in a "resonance" and a "phase space background."

With the understanding of this separation, the matrix B_{22} can be considered to represent the contributions of uncorrelated three-particle states to two-body reactions. What can we learn about them? From the field-theoretic side almost nothing is known about their structure. There is, however, one feature which may allow a treatment of these contributions. This is the notion that the matrices K_{23}, K_{33}, etc., have very high dimensions in actual cases. The angular momenta l, l' [see equations (10) and (11)] estimated from the available energy $\sqrt{\sigma}$, together with the number of possible final and initial states, *e.g.*, $N\pi K, N\pi\pi K, NK\bar{K}K, \Lambda KK$, etc., for, say, K^+p scattering at 3 GeV/c, give a very high number (around 100 in this case). Furthermore all matrix elements should be small and are of roughly the same order of magnitude. These facts lead us to suggest that a statistical treatment of these matrices should be possible.[9] More precisely, we assume that the values of the matrix elements of K_{23}, K_{32}, and K_{33} are randomly distributed around the mean value zero. From this we immediately reach two conclusions: B_{22} is purely imaginary and diagonal. Equation (16) tells us that Re B_{22} contains only odd powers of K_{33} and therefore vanishes because of our statistical assumption; the latter assumption also tells us that only diagonal elements, being sums of squares, survive.

If we make the additional assumption that *all* the elements of $K_{23}(K_{32})$ and K_{33} have the same probability distribution P_{23} and P_{33} respectively (P_{23} and P_{33} might well be different), we find that B_{22} is a (purely imaginary) multiple of the unit matrix

$$B_{22} = iCI \tag{19}$$

Our statistical assumptions lead therefore to the following result: The influence of all more-particle states on the two-body multichannel S-matrix is given by only one parameter for each total angular momentum J, isospin, *etc.*

For K_{22} we try the ansatz of Heitler's damping theory

$$K_{22} = \tfrac{1}{2} T_{22}^{\text{Born}} \tag{20}$$

This equation holds true in weak coupling limit. For strong coupling, where we actually need it, it has been shown that this ansatz bears quite important difficulties; it might give poles on the physical sheet, but it does not give a proper description of resonances, and it is not expected to predict very accurately the low partial waves. It fulfills however our condition (1) and certainly gives a unitary S-matrix. Now, if J_0 is not too high and if there are no resonances in the direct channel, then we do not need a correct prediction of the low partial waves, because these contribute only a small fraction of the differential cross section. What is, however, certainly remedied by the formulation (20) is the violent violation of unitarity displayed in Fig. 2. The factors 10. . . 70 involved there made the low partial wave contribution rather substantial. What we have done is just to propose two different origins of a damping of partial waves; the "self-damping" originating from (20) and an additional damping (19) produced by the experimentally important uncorrelated more-particle states. A necessary condition for the applicability of the model is, as stated above, that the energy region under consideration is far away from resonances in the direct channel, a condition which is fulfilled, say, for K^+p reactions at 3 GeV/c. Implications of the model for these reactions are presently being studied at CERN by H. Pilkuhn and myself.

REFERENCES

1. A. Martin, University of Washington Lectures, Spring, 1964.
2. R. Blankenbecler, Matscience Symposium, 1965.
3. M. J. Sopkovich, *Nuovo Cimento* **26**: 186 (1962).
4. K. Gottfried and J. D. Jackson, *Nuovo Cimento* **34**: 735 (1964).
5. L. Durand and Y. T. Chiu, *Phys. Rev. Letters* **12**: 399 (1964).
6. L. Lindenbaum, Dubna Conference, 1964.
7. K. Dietz and H. Pilkuhn, CERN preprint.
8. L. T. Cook and B. W. Lee, *Phys. Rev.* **27**: 283 (1962).
9. F. Dyson, *J. Math. Phys.* **3**: 140 (1962).

The Renormalizability of Higher Spin Theories

J. LUKIERSKI

UNIVERSITY OF WROCLAW
Wroclaw, Poland

1. THE PROJECTION OPERATOR ON THE DEFINITE SPIN SUBSPACE

We shall consider here the theories of elementary particles with spin $S \geqslant 1$. If we construct the theory of the particle with higher spin[†] and use the representation (i, j) of the Lorentz group, where $i \neq 0$ and $j \neq 0$, we have to introduce subsidiary conditions, singling out the components having definite spin. Very good examples are given by the Fierz theory of integer spin S, based on the representation $(S/2, S/2)$ and the Rarita–Schwinger theory of particles with half–integer spin $S + \frac{1}{2}$, based on the representation $\{(\frac{1}{2}, 0) \oplus (0, \frac{1}{2})\} \otimes (S/2, S/2)$. The components of the representation $(S/2, S/2)$ are described uniquely by the components of symmetric traceless tensor $U_{\mu_1 \mu_2 \ldots \mu_s}$:

$$\psi_{AB} = \tau_{AB}^{\mu_1 \ldots \mu_s} U_{\mu_1 \ldots \mu_s} \tag{1.1}$$

where $\mu_i = 1, 2, 3, 4$; $A, B = (S/2), (S/2) - 1, \ldots, -(S/2)$; and $\tau_{AB}^{\mu_1 \ldots \mu_s}$ is a generalization of well-known Pauli matrices σ_μ to arbitrary spin.[1]

ψ_{AB} contains, besides the representation D_s describing the S-spin particle, all other representations $D_{s'}$ with $s' < s$.[‡] We can

---

[†] *Higher spin* means equal to or greater than one.

[‡] This follows from the Clebsch-Gordan theorem:

$$D_{s/2} \otimes D_{s/2} = D_s \oplus D_{s-1} \oplus \ldots \oplus D_1 \oplus D_0$$

We recall that spin does not distinguish between undotted lower indices and dotted upper indices.

remove them if we consider a tensor $U_{\mu_1 \ldots \mu_s}$ traceless and symmetric, and satisfying the equation

$$\partial^\mu U_{\mu_1 \ldots \mu_s} = 0 \tag{1.2}$$

We shall introduce a tensor $\tilde{U}_{\mu_1 \ldots \mu_s}$ for which (1.2) is satisfied identically:

$$\psi_{AB}(s) = T_{AB}^{\mu_1 \ldots \mu_s} \tilde{U}_{\mu_1 \ldots \mu_s}$$

$$= T_{AB}^{\mu_1 \ldots \mu_s} \left(\frac{p_\mu}{p} \right) U_{\mu_1 \ldots \mu_s} \tag{1.3}$$

where

$$T_{AB}^{\mu_1 \ldots \mu_s} \left(\frac{p_\mu}{p} \right) = \tau_{AB}^{\nu_1 \ldots \nu_s} P_{\nu_1 \ldots \nu_s}^{\mu_1 \ldots \mu_s} \left(\frac{p_\mu}{p} \right) \tag{1.4}$$

and

$$p_\nu = i \partial_\nu \tag{1.5}$$

$$p^2 = p_\mu p^\mu = p^\mu p_\mu \tag{1.6}$$

$P_{\nu_1 \ldots \nu_s}^{\mu_1 \ldots \mu_s}$ is a covariant projection operator constructed from Kronecker deltas and the components of a unit vector p_μ/p. It is a homogeneous polynomial of order zero in p_ν with maximal number $2s$ of the terms p_μ/p. The coefficients of this polynomial we calculate assuming that

$$\partial^\mu P_{\mu \mu_2 \ldots \mu_s}^{\nu_1 \ldots \nu_s} = 0 \tag{1.7}$$

For $S = 1$, we get

$$P_\mu^\nu \left(\frac{p_\mu}{p} \right) = \delta_\mu^\nu - \frac{p_\mu p^\nu}{p^2} \tag{1.8}$$

and for $S = 2$

$$P_{\mu_1 \mu_2}^{\nu_1 \nu_2} \left(\frac{p_\mu}{p} \right) = \delta_{\mu_1}^{\nu_1} \delta_{\mu_2}^{\nu_2} - \frac{2}{3} \delta_{\mu_1}^{\nu_2} \delta_{\mu_2}^{\nu_1} - \frac{2}{3} \delta_{\mu_1 \mu_2} \delta^{\nu_1 \nu_2} - \frac{4}{9} \delta_{\mu_1 \mu_2} \frac{p^{\nu_1} p^{\nu_2}}{p^2} + \frac{2}{3} \delta_{\mu_1}^{\nu_1} \frac{p_{\mu_2} p^{\nu_2}}{p^2}$$

$$+ \frac{2}{3} \delta_{\mu_1}^{\nu_2} \frac{p_{\mu_2} p^{\nu_1}}{p^2} + \frac{2}{3} \delta_{\mu_2}^{\nu_1} \frac{p_{\mu_1} p^{\nu_2}}{p^2} + \frac{2}{3} \delta_{\mu_2}^{\nu_2} \frac{p_{\mu_1} p^{\nu_1}}{p^2} - \delta^{\nu_1 \nu_2} \frac{p_{\mu_1} p_{\mu_2}}{p^2} - \frac{8}{9} \frac{p_{\mu_1} p_{\mu_2} p^{\nu_1} p^{\nu_2}}{p^4} \tag{1.9}$$

One can check that $\psi_{AB}(s)$ in (1.3) has only $(2s + 1)$ independent components.

In the case of half-integer spin, we shall introduce two spinor-tensors τ^+ and τ^-. We have

$$\left\{\left(\tfrac{1}{2}, 0\right) \oplus \left(0, \tfrac{1}{2}\right)\right\} \otimes \left(\tfrac{s}{2}, \tfrac{s}{2}\right)$$

$$= \left(\tfrac{s+1}{2}, \tfrac{s}{2}\right) + \left(\tfrac{s}{2}, \tfrac{s+1}{2}\right) + \left(\tfrac{s-1}{2}, \tfrac{s}{2}\right) + \left(\tfrac{s}{2}, \tfrac{s-1}{2}\right) \quad (1.10)$$

If we denote

$$\chi^+_{AB} \sim \left(\tfrac{s+1}{2}, \tfrac{s}{2}\right) \qquad A = \tfrac{s+1}{2}, \tfrac{s-1}{2} \ldots - \tfrac{(s+1)}{2}$$

$$\chi^-_{BA} \sim \left(\tfrac{s}{2}, \tfrac{s+1}{2}\right) \qquad B = \tfrac{s}{2} \ldots - \tfrac{s}{2} \qquad (1.11)$$

we define τ^+, τ^- as follows:

$$\chi^{\pm}_{\bar{C}D} = \tau^{\pm \alpha \mu_1 \ldots \mu_s}_{\bar{C}D} \Phi_{\alpha \mu_1 \ldots \mu_s} \qquad (1.12)$$

where $\Phi_{\alpha \mu_1 \ldots \mu_s}$ is a symmetric traceless spinor tensor corresponding to representation (1.10).

If we wish to single out in the representations (1.7)

$$\left(\tfrac{s}{2} + \tfrac{\epsilon}{2}, \tfrac{s}{2} + \tfrac{1-\epsilon}{2}\right) \sim D_{s+1/2} \oplus D_{s-1/2} \oplus \ldots \oplus D_{1/2} \qquad (1.13)$$

(where $\epsilon = 0, 1$) only the highest spin representation $D_{s+1/2}$, we have to introduce $\Phi_{\alpha \mu_1 \ldots \mu_s}$, satisfying the following two conditions:

$$\partial^\mu \Phi_{\alpha \mu \mu_2 \ldots \mu_s} = 0$$
$$\gamma^{\mu}{}_\alpha{}^\beta \Phi_{\beta \mu \mu_2 \ldots \mu_s} = 0 \qquad (1.14)$$

We shall introduce $\widetilde{\Phi}_{\beta \mu_1 \ldots \mu_s}$, satisfying (1.14) as identities:

$$\chi^{\pm}_{\bar{C}D}(s + \tfrac{1}{2}) = \tau^{\pm \alpha \mu_1 \ldots \mu_s}_{\bar{C}D} \widetilde{\Phi}_{\alpha \mu_1 \ldots \mu_s}$$
$$= T^{\pm \alpha \mu_1 \ldots \mu_s}_{\bar{C}D} \Phi_{\alpha \mu_1 \ldots \mu_s} \qquad (1.15)$$

where

$$T^{\pm \alpha \mu_1 \ldots \mu_s}_{\bar{C}D} = \tau^{\pm \beta \nu_1 \ldots \nu_s}_{\bar{C}D} p^{\alpha \mu_1 \ldots \mu_s}_{\beta \nu_1 \ldots \nu_s}\left(\tfrac{p^\mu}{p}\right) \qquad (1.16)$$

The projection operator $P^{\alpha[\mu]}_{\beta[\nu]}$† satisfies the equalities (1.14) as identities:

$$\partial^\nu P^{\alpha \mu_1 \ldots \mu_s}_{\beta \nu \nu_2 \ldots \nu_s} \equiv 0$$
$$\gamma^{\nu}{}_\delta{}^\beta P^{\alpha \mu_1 \ldots \mu_s}_{\beta \nu \nu_2 \ldots \nu_s} \equiv 0 \qquad (1.17)$$

---

†We shall sometimes denote the set of indices $\mu_1 \ldots \mu_s$ by $[\mu]$ and $\nu_1 \ldots \nu_s$ by $[\nu]$.

and is symmetric and traceless in the tensor indices. For spin $s = \frac{3}{2}$, we have

$$P^{\alpha\mu}_{\beta\nu}\left(\frac{p_\mu}{p}\right) = \delta^\alpha_\beta \delta^\mu_\nu - \frac{1}{3}(\gamma_\nu \gamma^\mu)^\alpha_\beta$$

$$+ \frac{1}{3p^2}(\gamma_\nu{}^\delta_{,\beta} p^\mu - \gamma^\mu{}^\delta_{,\beta} p_\nu)(\gamma_{\sigma,}{}^\beta_\delta p^\sigma) - \frac{2}{3}\frac{p_\nu p^\mu}{p^2} \qquad (1.18)$$

One can calculate the formulas for $s = \frac{5}{2}, \frac{7}{2}$, etc., but the form of the coefficients is not known for the general case.

It is very important to mention that the projection operators satisfying (1.7) and (1.17), singling out the component with one spin value, are *completely independent* of the equations satisfied by the tensor (or tensor-spinor) representations. Only group theoretical considerations lead to the formulas (1.8) and (1.9) or (1.18). *We still have freedom in choosing the field equations.*

2. THE FIELD EQUATIONS: CONDITIONAL AND UNCONDITIONAL PROJECTION OPERATOR

We shall consider from now on mainly the integer spin fields. As we have mentioned, there is a freedom of choosing the equation for $U_{\mu_1\dots\mu_s}$. Generally, we have

$$U_{\mu_1\dots\mu_s} = \int_0^\infty dm\,\rho(m) U_{\mu_1\dots\mu_s}(m) \qquad (2.1)$$

where

$$p^2 U_{\mu_1\dots\mu_s}(m) = -m^2 U_{\mu_1\dots\mu_s}(m) \qquad (2.2)$$

The case of (2.2) corresponds to the free fields. *If we replace in the projection operator satisfying (1.7) all p^2 by $-m^2$, we get the conditional projection operator,* which restricts the representation (1.1) to the maximal spin subspace *only if the subsidiary components are free fields,* that is, equation (2.2) is satisfied.

Let us consider now quantum field theory. The commutator function

$$[U_{\mu_1\dots\mu_s}(x), U^{\nu_1\dots\nu_s}(y)] = F^{\nu_1\dots\nu_s}_{\mu_1\dots\mu_s}(x - y) \qquad (2.3)$$

describes the commutation relation of the quantized field with the definite spin S, if†

---

†The properties with respect to the second variable are obtained by

$$F_{\mu_1\dots\mu_s,\,\nu_1\dots\nu_s}(x - y) = -F_{\nu_1\dots\nu_s,\,\mu_1\dots\mu_s}(y - x) \qquad (2.3a)$$

$$F^{v_1...v_s}_{\mu_1...\mu_s}$$

is symmetric and traceless and

$$\partial^\mu F^{v_1...v_s}_{\mu\mu_2...\mu_s} = 0 \tag{2.4}$$

The projection operator obtained in Section 1 is symmetric. Therefore, $F^{\{v\}}_{\{\mu\}}$ can be written

$$F^{v_1...v_s}_{\mu_1...\mu_s}(x-y) = P^{v_1...v_s}_{\mu_1...\mu_s}\left(\frac{p_\mu}{p}\right)F(x-y) \tag{2.5}$$

where $F(x-y)$ is a scalar, odd function.[†] The formula (2.5) is again a result of purely group theoretical arguments. The field equation determines the equation for $F(x-y)$. If we look for the commutation relations of free particles with spin S, we shall have

$$(p^2+m^2)F(x-y)=0$$

or

$$F(x-y) \equiv i\Delta(m; x-y) \tag{2.6}$$

where $\Delta(m; x-y)$ is the well-known Jordan-Pauli commutator function for the free scalar field, and we can write (2.3) as follows:

$$[U^0_{\mu_1...\mu_s}(x), U^{0v_1...v_s}(y)] = iP^{v_1...v_s}_{\mu_1...\mu_s}\left(\frac{ip_\mu}{m}\right)\Delta(m; x-y) \tag{2.7}$$

We see that the unconditional projection operator is replaced by a conditional one. This is possible because of (2.6).

Now let us pose the following question: How do we construct a Lagrangian which leads to the commutation relation (2.7)? The answer is obtained by applying the theorem due to Takahashi and Umezawa.[2] The Lagrangian leading to (2.7) has a form

$$\mathcal{L}^0 = \frac{1}{2}U^0_{\mu_1...\mu_s}\Lambda^{\mu_1...\mu_s}_{v_1...v_s}\left(\frac{ip_\mu}{m}\right)U^{0v_1...v_s} \tag{2.8}$$

where

$$P^{v_1...v_s}_{\mu_1...\mu_s}\left(\frac{ip_\mu}{m}\right)\Lambda^{\rho_1...\rho_s}_{v_1...v_s}\left(\frac{ip_\mu}{m}\right) = -\delta^{\rho_1}_{\mu_1}\cdots\delta^{\rho_s}_{\mu_s}(p^2+m^2) \tag{2.9}$$

For example, because

$$\left(\delta^v_\mu+\frac{p_\mu p^v}{m^2}\right)(-(p^2+m^2)\delta^\rho_v+p_v p^\rho) = -\delta^\rho_\mu(p^2+m^2) \tag{2.10}$$

---

†From (2.3a) follows that
$$F(x-y) = -F(y-x)$$

for $S = 1$ we have

$$
\begin{aligned}
\mathscr{L}^0 &= \tfrac{1}{2}U_\mu(\square - m^2)U^\mu + \tfrac{1}{2}(U^\mu_\mu)^2 \\
&= -\tfrac{1}{2}[U_{\mu,\nu}U^{\mu,\nu} - (U^\mu_\mu)^2] - \tfrac{1}{2}m^2 U_\mu U^\mu \\
&= -\tfrac{1}{4}U_{\mu\nu}U^{\mu\nu} - \tfrac{1}{2}m^2 U_\mu U^\mu
\end{aligned} \tag{2.11}
$$

where

$$
U_{\mu\nu} = U_{\mu,\nu} - U_{\nu,\mu} \tag{2.12}
$$

Similarly, one can obtain Rarita-Schwinger Lagrangian for $S = \tfrac{3}{2}$, the Lagrangian for massive spin $S = 2$ particle, etc.

The point which we would like to bring about is that *if we do not substitute in the projection operator $ip \to m$, we are not able to solve equation* (2.9). From this follows that the unconditional projection operator cannot be present in any commutation relations obtained from the Lagrangian theory. The condition of solvability of (2.9) is nonsingularity of $P^{[\nu]}_{[\mu]}$. One can easily see that the conditional projection operator acts on the spaces with definite spins $S, S - 1$, ..., 2, 1, 0, which direct sum is described by the symmetric tensor $U_{\mu_1..\mu_s}$ as follows:

$$
\begin{aligned}
P_{ss} &= A_{ss}(p_\mu; m) \\
P_{s's''} &= A_{s's''}(p_\mu; m)(p^2 + m^2)
\end{aligned} \tag{2.13}
$$

where $s', s'' < s$ and

$$
A_{ss}(p_\mu; m)_{p^2 = -m^2} \neq 0 \tag{2.14}
$$

If the field is free, all $P_{s's''}$ are equal to zero, but if we consider the equation (2.9) without making the assumption $p^2 + m^2 = 0$ the conditional projection operator is nonsingular. For unconditional projection operator we have for all values of P_μ,

$$
P_{s's''} = 0 \tag{2.15}
$$

and the inverse Λ in the space of all components of symmetric tensor $U_{\mu_1...\mu_s}$ does not exist.

Therefore, we have learned that the unconditional projection operator cannot be obtained by the choice of the Lagrangian.

In the case of half-integer spin, we have the following relation replacing (2.5):

$$
F^{\beta\nu_1...\nu_s}_{\alpha\mu_1...\mu_s}(x - y) = P^{\gamma\nu_1...\nu_s}_{\alpha\mu_1...\mu_s}\left(\frac{p_\mu}{p}\right)F^\beta_\gamma(x - y) \tag{2.16}
$$

where $F^\beta_\gamma(x - y)$ is a commutator function of the spin $\tfrac{1}{2}$ particle:

$$(ip^\mu \gamma_\mu - m)_\alpha^{\ \gamma} F_\gamma^\beta (x - y) = 0 \tag{2.17}$$

It is worth mentioning that for $S = \tfrac{3}{2}$ we have

$$(p_\delta \gamma^\delta)_\alpha^\beta P_{\beta\nu}^{\delta\mu}\left(\frac{p_\mu}{p}\right) - P_{\alpha\nu}^{\beta\mu}\left(\frac{p_\mu}{p}\right)(p_\delta \gamma^\delta)_\beta^\delta = 0 \tag{2.18}$$

We see from (2.18) that the projection operator for $S = \tfrac{3}{2}$ commutes with the Dirac operator. It would be interesting to prove this property for arbitrary half-integer.

The conditional projection operator is obtained in (2.16) by putting $p^2 = -m^2$ or $p = \pm im$.

3. THE PROPAGATOR: NONCOVARIANT AND COVARIANT THEORY

If we have the commutator (2.7), we can calculate the function described by the time-ordered product of two $U_{\mu_1...\mu_s}$. We have

$$\langle 0| T U_{\mu_1...\mu_s}^0(x) U^{0\nu_1...\nu_s}(y)|0\rangle = \tfrac{1}{2} i F_{\mu_1...\mu_s}^{(1)\nu_1...\nu_s}(x - y)$$
$$- \tfrac{1}{2}\epsilon(x_0) F_{\mu_1...\mu_s}^{\nu_1...\nu_s}(x - y) \tag{3.1}$$

where $\epsilon(x_0)$ is a sign function, and

$$F_{\mu_1...\mu_s}^{(1)\nu_1...\nu_s}(x - y) = P_{\mu_1...\mu_s}^{\nu_1...\nu_s}\left(\frac{ip_\mu}{m}\right)\Delta^{(1)}(m; x - y) \tag{3.2}$$

$\Delta^{(1)}$ satisfies the Klein–Gordon equation and corresponds to the path of integration of Fourier integral in complex k_0-plane encircling clockwise the positive frequencies pole and anticlockwise the negative frequencies pole. There are drawn in Fig. 1 the paths of integration for Δ, $\Delta^{(1)}$, and Δ^c. The formula (3.1) can be obtained if

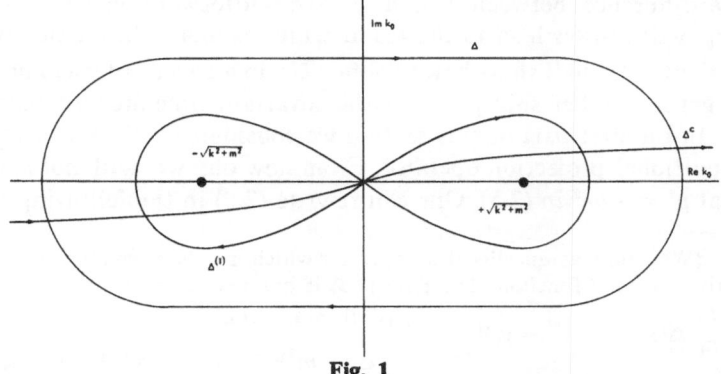

Fig. 1

we assume the paths of integration in F, $F^{(1)}$, and G to be the same as in Δ, $\Delta^{(1)}$, and Δ^c, respectively. But in order to get from (3.1) a relativistically covariant expression we have to add to (3.1) a quasilocal polynomial[†]

$$-\frac{i}{2}\left[P^{\nu_1\ldots\nu_s}_{\mu_1\ldots\mu_s}\left(\frac{ip_\mu}{m}\right),\,\epsilon(x_0)\right]\Delta(m;\,x) \tag{3.3}$$

and we get the following formula for the covariant causal propagator:

$$G^{\nu_1\ldots\nu_s}_{\mu_1\ldots\mu_s}(x-y) = iP^{\nu_1\ldots\nu_s}_{\mu_1\ldots\mu_s}(ip_\mu)\Delta^c(m;\,x-y)$$

$$= \langle 0|TU^0_{\mu_1\ldots\mu_s}(x)U^{0\nu_1\ldots\nu_s}(y)|0\rangle - \frac{i}{2}\left[P^{[\nu]}_{[\mu]}\left(\frac{ip_\mu}{m}\right),\,\epsilon(x_0-y_0)\right]\Delta(m;x-y)$$

$$\tag{3.4}$$

where

$$\Delta^c = \tfrac{1}{2}i\Delta^{(1)} - \tfrac{1}{2}\epsilon\Delta$$

The situation that the time-ordered product is not covariant is completely unsatisfactory. It seems that the formula for the commutator function F and the function $F^{(1)}$ should be changed to make the equality

$$\widetilde{G}^{\nu_1\ldots\nu_s}_{\mu_1\ldots\mu_s}(x-y) = \langle 0|T\widetilde{U}^0_{\mu_1\ldots\mu_s}(x)\widetilde{U}^{0\nu_1\ldots\nu_s}(y)|0\rangle$$

$$= \tfrac{1}{2}\widetilde{F}^{(1)\nu_1\ldots\nu_s}_{\mu_1\ldots\mu_s}(x-y) + i\epsilon(x_0)\widetilde{F}^{\nu_1\ldots\nu_s}_{\mu_1\ldots\mu_s}(x-y) \tag{3.5}$$

covariant. This will correspond to the modification of the operator $U^0_{[\mu]}$:

$$\widetilde{U}^0_{\mu_1\ldots\mu_s}(x) = U^0_{\mu_1\ldots\mu_s}(x) + \delta U_{\mu_1\ldots\mu_s}(x) \tag{3.6}$$

where the correction $\delta U_{[\mu]}$ commutes with $U_{[\mu]}$ and is responsible for the difference between F and $\widetilde{F}$. Straightforward calculations of $\delta U_{[\mu]}$ will always lead to the result which is not relativistically covariant. We shall show how to define $\widetilde{U}_{[\mu]}$ in a covariant way and how to get for higher spin particles the covariant time-ordered product.

In the first part of this section we considered a theory with the conditional projection operator. From now on, we will not assume that $p^2 = -m^2$ in (2.5). One can rewrite (2.5) in the following way:

---

†We define as quasilocal an operator which is a linear combination of the derivatives of δ-function. The term (3.3) is quasilocal, becuse

$$\frac{\partial^n}{\partial t^n}\Delta(\mathbf{x}-\mathbf{y},\,x_0-y_0)\bigg|_{x_0=y_0} = \begin{cases} 0 & \text{if } n \text{ is even} \\ (\Delta-m^2)^{(n-1)/2}\delta(\mathbf{x}-\mathbf{y}) & \text{if } n \text{ is odd} \end{cases}$$

$$\widetilde{F}^{\nu_1\dots\nu_s}_{\mu_1\dots\mu_s}(x-y) = D^{\nu_1\dots\nu_s}_{\mu_1\dots\mu_s}(p_\mu)F^s(x-y) \tag{3.7}$$

where $D^{[\nu]}_{[\mu]}$ is a homogeneous polynomial of order $2s$ in p_μ:

$$D^{\nu_1\dots\nu_s}_{\mu_1\dots\mu_s}(p_\mu) = (-1)^s p^{2s} P^{\nu_1\dots\nu_s}_{\mu_1\dots\mu_s}\left(\frac{p_\mu}{p}\right) \tag{3.8}$$

and

$$F^{(s)}(x-y) = \frac{(-1)^s}{p^{2s}}F(x-y) \tag{3.9}$$

Now if F satisfies (2.6), the corresponding equation for $F^{(s)}$ looks as follows:

$$p^{2s}(p^2+m^2)F^{(s)}(x-y) = (\Box - m^2)\Box^s F^{(s)}(x-y) = 0 \tag{3.10}$$

Let us define $F^{(s)}$ by a path of integration as in Fig. 2. The poles in the points $k_0 = \pm k$ are of order s. Because we have an identity

$$\frac{1}{\Box^s(\Box - m^2)} = \frac{1}{m^{2s}(\Box - m^2)} - \sum_{i=1}^{i=s} \frac{1}{\Box^i m^{2(s-i+1)}} \tag{3.11}$$

we can write

$$F^{(s)}(x-y) = \frac{1}{m^{2s}}i\Delta(m^2; x-y) - \frac{1}{m^2}\sum_{i=1}^{i=s}\frac{1}{m^{2(s-i)}}i\Delta^i(0; x-y) \tag{3.12}$$

where $\Delta^i(0; x-y)$ is a commutator function for the field equation

$$(-p^2)^i\varphi = \Box^i\varphi = 0 \tag{3.13}$$

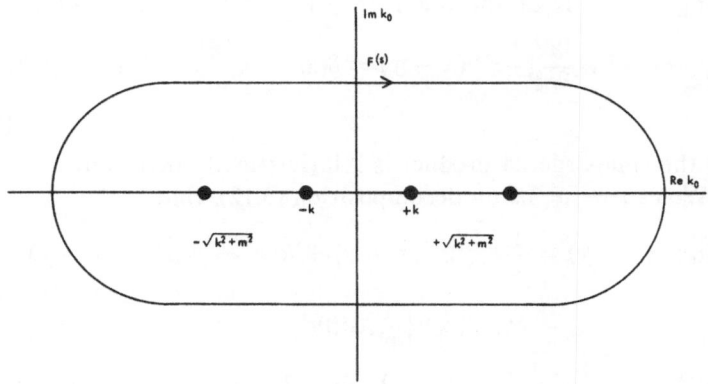

Fig. 2

and can be represented by the following formula:

$$\Delta^k(0; z) = \frac{\epsilon(z^0)}{8\pi k!}\kappa_{k-1}\left(\frac{z^2}{4}\right) \tag{3.14}$$

where $z^2 = z_i^2 - z_0^2$ and κ_n is a distribution defined as follows:

$$\kappa_{-n}(x) = \delta^{(n)}(x)$$
$$\kappa_1(x) = \theta(x)$$
$$\kappa_n(x) = \theta(x)\frac{x^{n-1}}{n!} \tag{3.15}$$

where $n \geqslant 2$.

The meaning of the distributions κ_n with $n \geqslant 1$ can be illustrated by the relation

$$\int \kappa_n(x)f^{(n)}(x)dx = (-1)^n f(0) \tag{3.16}$$

Equations (3.7) and (2.7) give a completely different result. But it is easy to see that (3.7) allows us to define the covariant propagator as a time-ordered product. This follows from the formula

$$\frac{\partial^m}{\partial t^m}F^{(s)}(\mathbf{x}, t)\Big|_{t=0} = \begin{cases} 0 & m \leqslant 2s - 1 \\ \delta(\mathbf{x}) & m = 2s \end{cases} \tag{3.17}$$

The term corresponding to (3.3) is now

$$i[D_{\mu_1...\mu_s}^{\nu_1...\nu_s}(p_\mu), \epsilon(x_0)]F^{(s)}(x - y) \tag{3.18}$$

and the highest time derivative in $D_{[\mu]}^{[\nu]}$ is $2s$. Because at least one time derivative has to act on $\epsilon(x_0)$, the highest possible derivative acting on $F^{(s)}$ is of the order $2s - 1$. According to (3.18), we have

$$\frac{d}{dx_0}\epsilon(x_0 - y_0)\frac{\partial^{2s-1}}{\partial x_0^{2s-1}}F^{(s)}(x - y) = \delta(x_0 - y_0)\frac{\partial^{2s-1}}{\partial x_0^{2s-1}}F^{(s)}(x - y, 0) = 0 \tag{3.20}$$

and the time-ordered product is relativistically invariant.

We can write, using decomposition (3.12), that

$$\tilde{G}_{\mu_1...\mu_s}^{\nu_1...\nu_s}(x - y) = \frac{1}{2}\tilde{F}_{\mu_1...\mu_s}^{(1)\nu_1...\nu_s}(x - y) + i\epsilon(x_0 - y_0)\tilde{F}_{\mu_1...\mu_s}^{\nu_1...\nu_s}(x - y)$$

$$= D_{\mu_1...\mu_s}^{\nu_1...\nu_s}(p_\mu)i\left\{\frac{1}{m^{2s}}\Delta^c(m^2; x - y)\right.$$

$$\left. + \frac{1}{m^2}\sum_{i=1}^{i=s}\frac{1}{m^{2(j-i)}}\Delta^{(i)c}(0; x - y)\right\} \tag{3.21}$$

where

$$\Delta^{(J)c}(0, x - y) = \frac{1}{(2\pi)^4} \lim_{\epsilon \to 0} \int d^4k \frac{e^{ik(x-y)}}{(k^2 - i\epsilon)^3} \quad (3.21a)$$

We can rewrite (3.11) using the fact that each term of the projection polynomial D has a form $D\binom{\beta_1 \ldots \beta_i}{\alpha_1 \ldots \alpha_i}; p_\mu)p^{2i}$ where

$$D_{[\mu]}^{[\nu]}\binom{\beta_1 \ldots \beta_i}{\alpha_1 \ldots \alpha_i}; p_\mu) = \prod_{j=1}^{j=i} \delta_{\mu\alpha_j}^{\nu\beta_j} p_{\mu\alpha_{i+1}} \ldots p_{\mu\alpha_s} p^{\nu\beta_{i+1}} \ldots p_{\nu\beta_s} \quad (3.22)$$

and $(\alpha_1 \ldots \alpha_i)$, $(\beta_1 \ldots \beta_i)$ are two sequences with $i \leqslant 2s$. We have

$$D_{\mu_1 \ldots \mu_s}^{\nu_1 \ldots \nu_s}(p_\mu) = \sum_{i=0}^{2s} D_{[\mu]}^{[\nu]}\binom{\beta_1 \ldots \beta_i}{\alpha_1 \ldots \alpha_i}; p_\mu)p^{2i} \quad (3.23)$$

$$D_{\mu_1 \ldots \mu_s}^{\nu_1 \ldots \nu_s}(p_\mu)\Delta^{(J)c}(x - y) = \sum_{\substack{i=0 \\ (\alpha_1 \ldots \alpha_i) \\ (\beta_1 \ldots \beta_i)}}^{2s} D_{[\mu]}^{[\nu]}\binom{\beta_1 \ldots \beta_i}{\alpha_1 \ldots \alpha_i})\Delta^{(J-i)c}(x - y)$$

$$= \sum_{i=0}^{2s} D_{[\mu]}^{[\nu]}\binom{\beta_1 \ldots \beta_i}{\alpha_1 \ldots \alpha_i}; p_\mu)\left\{\frac{1}{m^{2(s-i)}}\Delta^c(m^2; x - y)\right.$$

$$\left. - \frac{1}{m^2}\sum_{j=1}^{j=i}\frac{1}{m^{2(j-i)}}\Delta^{(J)c}(0; x - y)\right\} \quad (3.24)$$

But

$$\sum_{\substack{i=0 \\ (\alpha_1 \ldots \alpha_i) \\ (\beta_1 \ldots \beta_i)}}^{2s} D_{[\mu]}^{[\nu]}\binom{\beta_1 \ldots \beta_i}{\alpha_1 \ldots \alpha_i}; p_\mu)\frac{1}{m^{2(s-i)}} = p_{[\mu]}^{[\nu]}\left(\frac{ip_\mu}{m}\right) \quad (3.25)$$

and we have finally

$$\widetilde{G}_{\mu_1 \ldots \mu_s}^{\nu_1 \ldots \nu_s}(x - y) = G_{\mu_1 \ldots \mu_s}^{\nu_1 \ldots \nu_s}(x - y)$$

$$- \frac{1}{m^2}\sum_{i=1}^{2s} m^{2i} D_{[\mu]}^{[\nu]}\binom{\beta_1 \ldots \beta_i}{\alpha_1 \ldots \alpha_i}\sum_{j=1}^{j=i}\frac{1}{m^{2j}}\Delta^{(J)(c)}(0; x - y) \quad (3.26)$$

We see therefore that in the formulation with unconditional projection operator we have to add a set of S massless fields, whose metric is negative, but we gain two very important properties of the theory: (a) invariance of the T-product; and (b) much less stringent renormalizability requirements. In the next section, we shall discuss the second point.

4. THE RENORMALIZABILITY PROBLEM

Let us assume that the higher spin particles are described by the propagator (3.4). The scattering of higher spin particles with the massive scalar bosons, photons, and fermions with spin $\frac{1}{2}$ can be described by a set of all Feynman graphs with the vertices having not more than f fermion lines, b scalar bosons, e photon lines, and h lines propagating the particles with spin $s \geqslant 1$. If we further introduce the parameter d, describing the number of derivatives, we shall obtain, using Dyson's criteria of renormalizability of ultraviolet divergences, the following condition on the parameters:

$$\tfrac{3}{2}f + (1 + s)h + b + e + d \leqslant 4 \tag{4.1}$$

If we introduce two parameters

$$S = \tfrac{1}{2}f + sh$$
$$N = f + h + b + e \tag{4.2}$$

where S is a sum of spins of all massive particles taking part in the interaction, and N is the number of particles, we can write (4.1) as follows:[†]

$$S + N \leqslant 4 - d \tag{4.3}$$

We see that in the usual formalism *spin counts in the same way as the number of fields*, and because for local theory, with $d \geqslant 0$ at least $N = 3$ and $S \leqslant \tfrac{1}{2}$,[‡] we obtain the result that all theories of higher spin with the conventional propagator (3.4) are nonrenormalizable. The situation is completely changed if we introduce the propagator (3.26). We have, instead of (4.3),

$$\tfrac{1}{2}F + N \leqslant 4 - d \tag{4.4}$$

where F describes the number of fermions, and we get the result that:

If $N = 3$, there are two classes of renormalizable interactions: (a) all interactions without derivatives involving not more than two fermions, and (b) all interactions with one derivative if the fermions are not present.

If $N = 4$, all interactions are renormalizable that do not involve fermions and derivatives.

We see that there is a difference between fermions and bosons. This difference follows from the fact that all fermions, we assume, satisfy the Dirac equation, and all bosons only the Klein–Gordon equation. The unconditional projection operator does not change

---

[†]We can also include in N massless higher spin particles.

[‡]From (4.3) it follows that $S < 1$, but relativistic invariance implies that for $S = 1$, $d = 1$.

the degree at infinity of the Fourier component of the propagator. Finally, all propagators of bosons behave in momentum space k_μ for $k \to \infty$ as $1/k^2$, and all fermion propagators as $1/k$.

As we can see from the above discussion, the two theories, one with conditional and the other with unconditional projection operator, are essentially different. The free propagators of both theories are connected by the formula (3.21). Now we shall introduce two classes of interactions: (a) those which preserve formula (3.26); that is, the additional S massless fields, contributing to $\widetilde{G}$, remain free and do not interact; and (b) the interactions which do not preserve formula (3.26).

These two classes of interaction are easy to distinguish if we assume the interaction Lagrangian in the form

$$L^{\text{int}} = g U_{\mu_1 \ldots \mu_s} T^{\mu_1 \ldots \mu_s} \tag{4.5}$$

where $T^{\mu_1 \ldots \mu_s}$ is a general current not depending on $U_{\mu_1 \ldots \mu_s}$. If the theory is of the first type, we have the following conservation laws satisfied:

$$T_{\mu_1 \ldots \mu_s} \text{ is symmetrical} \tag{4.6}$$
$$T^\mu_{\mu \mu_3 \ldots \mu_s} = 0$$
$$\partial^\mu T_{\mu \mu_2 \ldots \mu_s} = 0$$

They correspond exactly to the subsidiary conditions on $U_{\mu_1 \ldots \mu_s}$.

As we have already mentioned, we *cannot* construct the theory with an unconditional projection operator, leading to renormalizable theory, using usual Lagrangian approach. The situation is similar to that of the Landau gauge in electrodynamics, which cannot be derived from the Lagrangian employing only four potentials A_μ.[†] However one can construct the theory of the electromagnetic field in Landau gauge, if we add to A_μ a fifth field λ and impose a subsidiary condition on A_μ and λ (see Appendix). A similar method can be employed in order to obtain the theory of higher spin field with unconditional projection operator. We introduce additional fields $B_1 \ldots B_s$, which are responsible for the difference between the propagator $\widetilde{G}$ and G [see (3.26)].

In order to illustrate the method of constructing theory which

---

[†]We recall, that the gauge-invariant Lagrangian $-\frac{1}{4} F_{\mu\nu} F^{\mu\nu}$ leads to the commutation relations of the potentials with *arbitrary* longitudinal part $\alpha(x - y)$:

$$[A_\mu(x), A_\nu(y)] = \left(\delta_{\mu\nu} - \frac{\partial_\mu \partial_\nu}{\Box}\right) \Delta(x - y) + \frac{\partial_\mu \partial_\nu}{\Box} \alpha(x - y)$$

leads to the propagator $\tilde{G}$, we shall investigate in the next section the simplest higher spin field, a neutral massive vector meson. It is hoped that this example will explain also the physical meaning of the theory with unconditional spin projection operator.

5. AN EXAMPLE: NEUTRAL VECTOR MESON THEORY. EXTENDED STUECKELBERG FORMALISM

We shall now apply the method, used in the Appendix, to the massive spin-1 particle. For simplicity, we assume that the field is real.†

It is known that the usual theory of spin-1 particles (with the conditional projection operator) is described by the Lagrangian (2. 11), and the field equations

$$\partial^\mu U_{\mu\nu} - m^2 U_\nu = j_\nu \tag{5.1}$$

can be obtained in a way first proposed by Stueckelberg.[5] We start with the free field case. We introduce the five fields A_μ, B governed by the following Lagrangian

$$L_0 = -\frac{1}{2}(\partial_\mu A_\nu)^2 - \frac{m^2}{2}A_\mu A^\mu - \frac{1}{2}(\partial_\mu B)^2 - \frac{m^2}{2}BB \tag{5.2}$$

we have

$$(\Box - m^2)A_\mu = 0 \tag{5.3a}$$

$$(\Box - m^2)B = 0 \tag{5.3b}$$

and

$$[A_\mu(x), A_\nu(y)] = i\delta_{\mu\nu}\Delta(m; x - y) \tag{5.4a}$$

$$[B(x), B(y)] = i\Delta(m; x - y) \tag{5.4b}$$

$$[A_\mu(x), B(y)] = 0 \tag{5.4c}$$

If we introduce

$$U_\mu = A_\mu + \frac{1}{m}B_{,\mu} \tag{5.5}$$

we can write (5.1) as follows:

$$L_0 = -\frac{1}{2}(\partial_\mu U_\nu)^2 - \frac{1}{2}(\partial_\mu U^\mu)^2 - \frac{m^2}{2}U_\mu U^\mu - \frac{1}{2}\chi^2 \tag{5.5a}$$

---

†If U_μ is complex, one has to consider a sum of two Lagrangians for real vector fields $U_\mu^{(1)}$, $U_\mu^{(2)}$ where $U_\mu = U_\mu^{(1)} + iU_\mu^{(2)}$.

where

$$\chi = \partial_\mu A^\mu + mB \tag{5.6}$$

If we assume

$$\chi|ph\rangle = 0 \tag{5.7}$$

the Lagrangians (5.1) and (2.11) on the physical space are the same. The subsidiary condition (5.7) can be written as follows:

$$U_\mu^\mu|ph\rangle = \frac{1}{m}(\square - m^2)B|ph\rangle \tag{5.8}$$

and (5.7) is equivalent to

$$U_\mu^\mu|ph\rangle 0 \tag{5.9}$$

provided the field B is free.

In order to preserve the condition (5.6), we must introduce the interaction as follows:

$$L_{\text{Int}}(U_\mu; \Phi_A) = L_{\text{Int}}\left(A_\mu + \frac{1}{m}B_{,\mu}; \Phi_A\right) \tag{5.10}$$

where Φ_A are the fields interacting with U_μ and the equation for B takes a form

$$(\square - m^2)B = -\frac{1}{m}\partial_\mu\frac{\delta L_{\text{Int}}}{\delta U_\mu} = -\frac{1}{m}j_\mu^\mu \tag{5.11}$$

We see that if the current

$$j_\mu = \frac{\delta L}{\delta U_\mu} \tag{5.12}$$

is not conserved, the subsidiary condition (5.9) is violated.

The field U_μ satisfies the commutation relations with the conditional projection operator

$$[U_\mu(x), U_\nu(y)] = i\left(\delta_{\mu\nu} - \frac{\partial_\mu\partial_\nu}{m^2}\right)\Delta(m; x - y) \tag{5.13}$$

The commutation relations with the unconditional projection operator can be written as follows:

$$[\tilde{U}_\mu(x), \tilde{U}_\nu(y)] = i\left(\delta_{\mu\nu} - \frac{\partial_\mu\partial_\nu}{m^2}\right)\Delta(m; x - y) + i\frac{\partial_\mu\partial_\nu}{m^2}\Delta(0; x - y)$$

$$= i[U_\mu(x), U_\nu(y)] + i\frac{\partial_\mu\partial_\nu}{m^2}\Delta(0; x - y) \tag{5.14}$$

because

$$\frac{1}{p^2(p^2 + m^2)} = -\frac{1}{m^2}\frac{1}{p^2 + m^2} + \frac{1}{m^2}\frac{1}{p^2} \tag{5.15}$$

or

$$\frac{1}{\Box(\Box - m^2)} = \frac{1}{m^2}\frac{1}{\Box - m^2} - \frac{1}{m^2}\frac{1}{\Box} \qquad (5.16)$$

We see therefore that in order to obtain the free field commutation relations with unconditional projection operator *we have to add to the Lagrangian (5.1) a second scalar field B'*, massless and with negative energy. We shall call theory so modified, an *extended Stueckelberg formalism*. We have for this theory the following free Lagrangian:

$$\widetilde{L}_0 = L_0 + \tfrac{1}{2}(\partial_\mu B')^2 \qquad (5.17)$$

and we have, besides the equations in (5.3) and (5.4), the following equations of motion and commutation relations:

$$\Box B' = 0 \qquad (5.3c)$$

$$[B'(x), B'(y)] = -i\Delta(0; x - y) \qquad (5.4d)$$

$$[B'(x), A_\mu(y)] = [B'(x), B(y)] = 0 \qquad (5.4e)$$

Now we show that if we impose the second subsidiary condition

$$\widetilde{\chi}|ph\rangle = (mU'' - B')|ph\rangle = 0 \qquad (5.18)$$

where

$$U'' = \frac{1}{\Box}\partial_\mu U^\mu \qquad (5.19)$$

the second compensating field B' eliminates the longitudinal part of U_μ from the mass term.

We introduce

$$\widetilde{U}_\mu = U_\mu - \frac{1}{m}B'_{,\mu} = A_\mu + \frac{1}{m}B_{,\mu} - \frac{1}{m}B'_{,\mu} \qquad (5.20)$$

We can write now (5.18) as follows:

$$m\widetilde{U}''|ph\rangle = (mU'' - B')|ph\rangle = 0 \qquad (5.21a)$$

or

$$\widetilde{U}_\mu|ph\rangle = (\widetilde{U}_\mu^\perp + \partial_\mu U'')|ph\rangle = \widetilde{U}_\mu^\perp|ph\rangle \qquad (5.21b)$$

We can express $\widetilde{L}'$ by $\widetilde{U}_\mu$ as follows:

$$\widetilde{L}_0 = -\tfrac{1}{2}\{(\partial_\mu \widetilde{U}_\nu)^2 - (\partial_\mu \widetilde{U}^\mu)^2 + m^2 \widetilde{U}_\mu \widetilde{U}^\mu\} - \tfrac{1}{2}\chi^2 - B'\Box\widetilde{U}'' \qquad (5.22)$$

In order to preserve the subsidiary condition (5.18) also in the presence of the interaction, we have to introduce the interaction

Lagrangian (5.10) with U_μ replaced by $\widetilde{U}_\mu$:

$$\widetilde{L}^{\text{int}}(U_\mu, B'; \Phi_A) = L_{\text{int}}(\widetilde{U}_\mu; \Phi_A)$$

$$= L_{\text{int}}\left(U_\mu - \frac{1}{m}B_{,\mu}; \Phi_A\right)$$

$$= L_{\text{int}}\left(A_\mu + \frac{1}{m}B_{,\mu} - \frac{1}{m}B'_\mu; \Phi_A\right) \tag{5.23}$$

The equations of motion which follow from (5.22) and (5.23) are

$$\delta\widetilde{U}_\mu: \ \partial^\mu \widetilde{U}_{\mu\nu} - m^2 \widetilde{U}_\nu + \chi_{,\nu} + B'_{,\nu} = j_\nu(\widetilde{U}_\mu; \Phi_A) \tag{5.24a}$$

$$\delta B: \ \chi = 0 \tag{5.24b}$$

$$\delta B': \ \widetilde{U}^\mu_\mu = 0 \tag{5.24c}$$

The subsidiary conditions (5.7) and (5.18) are obviously consistent with the equations even if

$$j^\mu_\mu \neq 0 \tag{5.25}$$

We have from (5.24) that

$$\square B' = j^\nu_\nu(\widetilde{U}_\mu; \Phi_A) \tag{5.26}$$

or, using (5.21b),

$$\square B'|ph\rangle = j^\nu_\nu(\widetilde{U}^\perp_\mu, \Phi_A)|ph\rangle \tag{5.27}$$

Substracting the equation for B' from (5.24a), we get

$$(\partial^\mu \widetilde{U}_{\mu\nu} - m^2 \widetilde{U}^\perp_\nu)|ph\rangle = \left(\delta_{\mu\nu} - \frac{\partial_\mu \partial_\nu}{\square}\right) j^\mu(\widetilde{U}^\perp_\mu; \Phi_A)|ph\rangle \tag{5.28}$$

This is the equation of motion for the spin-1 field. The projection operator, which assures the transversality of $\widetilde{U}^\perp_\nu$ even if $j^\mu_\mu \neq 0$ tells us that the equation (5.28) can be obtained in a much simpler way: by projecting on the spin-1 subspace the equation, obtained in usual spin-1 theory [from Lagrangian (2.11)]. This was done several years ago by Glashow,[3] who did not refer to any compensating field B'. However, our method has an advantage, which can be seen if we look at the interaction Lagrangian (5.23). Let us introduce the field $U^\perp_\mu$ which is purely transversal; that is, it satisfies the relation

$$\left(\delta^\nu_\mu - \frac{\partial_\mu \partial^\nu}{\square}\right) U^\perp_\nu \equiv U^\perp_\mu \tag{5.29}$$

We can then write

$$L^{\text{int}}(U^\perp_\mu; \Phi_A) \equiv L_{\text{int}}\left[\left(\delta^\nu_\mu - \frac{\partial_\mu \partial^\nu}{\square}\right) U^\perp_\nu; \Phi_A\right] \tag{5.30}$$

and we shall get

$$\frac{\delta L^{\text{int}}}{\delta U_\mu^\perp} \equiv \left(\delta_\mu^\nu - \frac{\partial_\mu \partial^\nu}{\Box}\right)\frac{\delta L^{\text{int}}}{\partial U_\nu^\perp} \tag{5.31}$$

Therefore, if the vector meson is purely transversal, the presence of the projection operator in (5.28) expresses only the fact that this equation has been obtained by performing purely transversal variations $\delta U_\mu^\perp$. *The interaction Lagrangian remains the same in form as in usual theory, only U_μ is replaced by $U_\mu^\perp$ and the propagator (5.13) by (5.14) (see Section 4).* In Glashow's approach we calculate also

$$U'' = \frac{1}{\Box}U_\mu^\mu$$

from the equation obtained by taking the longitudinal part of the equations (5.1) of the usual theory

$$\Box U'' = -\frac{1}{m^2}j_\nu^\nu \tag{5.32}$$

and substitute back into (5.1). In modified Stueckelberg formalism the longitudinal component satisfying (5.32) is compensated by B' satisfying (5.26) [this follows from the subsidiary condition (5.18)].

At the end of this section, we shall mention the conditions for the equivalence of two presented theories of spin-1. If the difference is expressed by free quanta of the field B, one can remove it by shifting the scale of eigenvalues of the total energy-momentum tensor by the energy and momentum of all quanta of the field B'. The difference, however, becomes essential if the quanta B' interact with the field U_μ. As one can see from (5.26), the condition is

$$j_\mu^\mu = 0 \tag{5.33}$$

in accordance with the general rules presented in Section 4 [formulas (4.6)].

6. FINAL REMARKS

In this lecture, we have presented the *modification* of the theory of higher spin particles, which leads to much less stringent conditions for the renormalizability. We have discussed mostly the higher spin particles with integer spin and mass $m \neq 0$. The generalization of the results to the case of half-integer spin is obvious however more complicated. The projection operator becomes a function of p_μ and γ_μ matrices, which creates new calculational problems because of the noncommutativity of γ_μ matrices.

The case of $m = 0$ is completely different from the case of massive particles. For the massless particles, the projection operator has to be always unconditional, and every massless particle has a generalized Landau gauge, defined by the unconditional projection operator and the commutator function (2.5). Because of this situation, it was felt long ago that all massless particles have the propagators behaving like $1/k^2$ for large momentum.

In order, however, to obtain the theory of higher spin particles, we should consider for spin S all independent solutions of the equation

$$\Box^s \varphi_\lambda^{(s)} = 0$$

What is very interesting is that the passage $m^2 \to 0$ does not present any difficulties and can be made in the commutator function or in the propagator. We have, for example, for the propagator of vector meson the following limit:

$$\Delta_{\mu\nu}^c(0; x - y) = \lim_{m^2 \to 0} \Delta_{\mu\nu}^c(m; x - y)$$

$$\Delta_{\mu\nu}^c(m; x - y) = \delta_{\mu\nu}\frac{1}{\Box - m^2 - i\epsilon} - \partial_\mu\partial_\nu\frac{1}{(\Box - i\epsilon')(\Box - m^2 - i\epsilon)}$$

and for $\epsilon' \to 0$ this limit gives the theory of massless vector meson (electrodynamics in the Landau gauge).

APPENDIX: LAGRANGIAN FORMULATION OF THE ELECTRODYNAMICS IN THE LANDAU GAUGE

Let us write the usual Lagrangian for massless spin-1 particle.

$$L_0 = -\tfrac{1}{2}(\partial_\mu A_\nu)^2 = -\tfrac{1}{4}F_{\mu\nu}F^{\mu\nu} - \tfrac{1}{2}(A_\mu^\mu)^2 \tag{A.1}$$

Now we shall add to (A.1) the free Lagrangian of a certain massless field λ:

$$L_0 \to L_0^\lambda = L_0 + \tfrac{1}{2}\lambda\Box^2\lambda \tag{A.2}$$

Let us introduce the following subsidiary condition on physical states $|ph\rangle$:

$$\chi|ph\rangle = \left(\lambda - \frac{A_\mu^\mu}{\Box}\right)|ph\rangle = (\lambda - A_\perp'')|ph\rangle = 0 \tag{A.3}$$

where

$$A'' = \frac{1}{\Box} A_\mu^\mu$$

Introducing now

$$\widetilde{A}_\mu = A_\mu - \partial_\mu \lambda \qquad (A.4)$$

we can write (A.2) as follows:

$$L_0^\lambda = -\tfrac{1}{4} F_{\mu\nu} F^{\mu\nu} - \tfrac{1}{2}\chi \Box^2 \chi + \lambda \Box^2 \lambda - A_\mu^\mu \Box \lambda$$

$$= -\tfrac{1}{4} F_{\mu\nu} F^{\mu\nu} - \tfrac{1}{2}\chi'^2 + \widetilde{A}_\mu^\mu \lambda' \qquad (A.5)$$

where

$$\chi' = \Box\chi \qquad \lambda' = \Box\lambda \qquad (A.6)$$

Because

$$\chi|ph\rangle = -\frac{\widetilde{A}^\mu_\mu}{\Box}|ph\rangle = 0 \qquad (A.7)$$

we have the result that

$$\widetilde{A}_\mu|ph\rangle = \widetilde{A}_\mu^\perp|ph\rangle \qquad (A.8)$$

and from (A.4) we get

$$[\widetilde{A}_\mu(x), \widetilde{A}_\nu(y)] = [A_\mu(x), A_\nu(y)] + \partial_\mu^x \partial_\nu^y [\lambda(x), \lambda(y)]$$

$$= \left(\delta_{\mu\nu} - \frac{\partial_\mu \partial_\nu}{\Box}\right) \Delta(0; x - y) \qquad (A.9)$$

which is exactly the Landau gauge.

It is worth mentioning that the subsidiary condition (A.3) [or equivalently (A.7) or (A.8)] is consistent with the commutation relations (A.9). Another feature of this formalism is that we can couple to A_μ also nonconserved current:

$$L_A^{\text{int}} = j_\mu A^\mu, \ j_\mu^\mu \neq 0 \qquad (A.10)$$

However, if we modify the theory by introducing the following coupling to the field λ:

$$L_\lambda^{\text{int}} = -j_\mu^\mu \lambda \qquad (A.11)$$

we can write

$$L_{\text{int}} = L_A^{\text{int}} + L_\lambda^{\text{int}} = j_\mu \widetilde{A}^\mu = j_\mu^\perp A_\mu^\perp + j_\mu^\mu(-\lambda + A'') \quad (A.12)$$

and the nonconserved part acts only in the nonphysical part of Hilbert space. Taking as independent variables $A_\mu^\perp$, A'' and λ, we get the following equations:

$$\partial^\mu F_{\mu\nu} = j_\nu^\perp$$
$$\Box^2 A'' = j_\mu^\mu \tag{A.13}$$
$$\Box^2 \lambda = j_\mu^\mu$$

and we get (even if there is an interaction with nonconserved current) the equation

$$\Box^2 \chi = 0 \tag{A.14}$$

consistent with (A.7).

REFERENCES

1. S. Weinberg, *Phys. Rev.* **133B** (1964).
2. Y. Takahashi and H. Umezawa, *Progr. Theoret. Phys.* **9**: 1 (1953).
3. S. Glashow, *Nuclear Phys.* **10**: 107 (1959).
4. H. Umezawa, *Quantum Field Theory*, Amsterdam, 1956.
5. Stueckelberg, *Acta Phys. Helv.* **11**: 299 (1938).

Group Representations for Complex Angular Momentum

K. VENKATESAN

MATSCIENCE
Madras, India

The notion of complex angular momentum has come to the fore in S-matrix theory since the work of Regge, who showed that for a complete description of potential scattering, an analytical continuation of the scattering amplitude into the complex angular momentum had to be made; but the idea is well-recognized in mathematics, where the Schläfli integral representation, with suitable cuts, is a solution of the Legendre equation for arbitrary values of l, the angular momentum quantum number. Attempts at extending the description in terms of Regge poles to the relativistic problem of elementary particle interactions have not been uniformly successful, though there are interactions (such as proton-proton or K^+ meson-proton scattering) in which the characteristic shrinking of the diffraction peak at high energies predicted by the Regge pole hypothesis is observed, and though there are situations such as that in the multiperipheral model of Amati *et al.*, where the summation over the ladder diagrams gives rise to a Regge pole behavior for the scattering amplitude. One would like to have such a behavior, because the Regge pole moving on its trajectory can describe a bound state or a resonance and also the high-energy behavior of cross sections—all that one could wish for from a scattering theory. It would therefore be interesting to see how a state with complex angular momentum can be described. States with nonphysical spins are, as stated above, like the unstable or virtual states, ephemeral, and wave functions describing them cannot be constructed without

91

violating the usual boundary conditions. We shall look at this problem from the point of view of representations for the various groups where a spin description enters.

The starting point would naturally be the rotation group or, to be more precise, the little group of the Poincaré group associated with a timelike vector. The irreducible finite-dimensional representations D^j of this group are labeled by integral or half-integral values for j. A general infinite-dimensional representation of the group in a reflexive Banach space may always be decomposed into the direct sum of representations each of which is a multiple of D^j. Thus, to introduce complex angular momentum into the group, one has to try other methods which, as we shall see, lead to structures more general than groups.

Beltrametti and Luzzatto use Weyl's method for constructing the representation of the rotation group, namely, to find out how a monomial constructed from two complex variables

$$f_m^\nu(u_1, u_2) = A_m^\nu u_1^{\nu+m} u_2^{\nu-m}$$

transforms when the basic components u_1 and u_2 of the two-dimensional spinor, $x = \begin{pmatrix} u_1 \\ u_2 \end{pmatrix}$ are transformed by the unitary unimodular matrix:

$$M = \begin{vmatrix} a & b \\ -b^* & a^* \end{vmatrix} = [a, b] \tag{1}$$

Here, ν is an arbitrary complex number, m can take integral or half-integral values, and

$$A_m^\nu = e^{im\pi}[\Gamma(\nu + m + 1)\Gamma(\nu - m + 1)]^{-1/2}$$

For $|a| > |b|$, and choosing $|u_1| = |u_2| = 1$, we have

$$f_m^\nu(M^{-1}x) = A_m^\nu(a^*u_1 - bu_2)^{\nu+m}(b^*u_1 + au_2)^{\nu-m}$$

$$= A_m^\nu \sum_{r=0}^{\infty} \sum_{s=0}^{\infty} (-1)^r a^{\nu-m-s}(a^*)^{\nu+m-r} b^r(b^*)^s$$

$$\times \frac{(\nu + m)\cdots(\nu - m)\cdots}{r!s!} u_1^{\nu+m-r+s} u_2^{\nu-m-r-s}$$

$$= \sum_{m'=-\infty}^{+\infty} B_{m',m}^\nu(M) f_{m'}^\nu(x) \tag{2}$$

where $(m' - m)$ is an integer.

The second step is possible only if $a \neq 0$. By comparison and a change of the order of summation we have two cases. For $m - m' \leq 0$,

$$B_{m',m}^{\nu}(M) = \left[\frac{\Gamma(\nu - m + 1)\Gamma(\nu + m' + 1)}{\Gamma(\nu + m + 1)\Gamma(\nu - m' + 1)}\right]^{1/2} \frac{(-1)^{m'-m}}{(m'-m)!}$$

$$\times a^{\nu-m'}(a^*)^{\nu+m}(b^*)^{m'-m} F\left(-\nu - m, -\nu + m'; m' - m + 1; -\left|\frac{b}{a}\right|^2\right)$$

$$= \left[\frac{\Gamma(\nu - m + 1)\Gamma(\nu + m' + 1)}{\Gamma(\nu + m + 1)\Gamma(\nu - m' + 1)}\right]^{1/2} \frac{(-1)^{m'-m}}{(m'-m)!}$$

$$\times a^{-m'-m}(b^*)^{m'-m} F(-\nu - m, \nu - m + 1; m' - m + 1; |b|^2)$$

$$(3)$$

The second step is obtained by using the identity

$$F(a_1, a_2, a_3; z) = (1 - z)^{-a_1} F\left(a_1, a_3 - a_2; a_3; \frac{z}{z - 1}\right) \qquad (4)$$

For $m - m' > 0$, we have

$$B_{m',m}^{\nu}(M) = \left[\frac{\Gamma(\nu + m + 1)\Gamma(\nu - m' + 1)}{\Gamma(\nu - m + 1)\Gamma(\nu + m' + 1)}\right]^{1/2} \frac{1}{(m - m')!}$$

$$\times a^{\nu-m}(a^*)^{\nu+m'} b^{m-m'} F\left(-\nu + m, -\nu - m'; m - m' + 1; -\left|\frac{b}{a}\right|^2\right)$$

$$= \left[\frac{\Gamma(\nu + m + 1)\Gamma(\nu - m' + 1)}{\Gamma(\nu - m + 1)\Gamma(\nu + m' + 1)}\right]^{1/2} \frac{1}{(m - m')!}$$

$$\times a^{-m-m'} b^{m-m'} F(-\nu - m'; \nu - m' + 1; m - m' + 1; |b|^2) \qquad (5)$$

For $|a| < |b|$, we have for $b \neq 0$ the following set of matrices into which the unitary unimodular group is mapped for complex values of ν: For $m + m' \leq 0$,

$$C_{m',m}^{\nu}(M) = \left[\frac{\Gamma(\nu - m + 1)\Gamma(\nu - m' + 1)}{\Gamma(\nu + m + 1)\Gamma(\nu + m' + 1)}\right]^{1/2} \frac{e^{i\pi(\nu+m')}}{(-m - m')!}$$

$$\times a^{-m-m'} b^{\nu+m}(b^*)^{\nu+m'} F\left(-\nu - m, -\nu - m'; -m - m' + 1; -\left|\frac{a}{b}\right|^2\right)$$

$$= \left[\frac{\Gamma(\nu - m + 1)\Gamma(\nu - m' + 1)}{\Gamma(\nu + m + 1)\Gamma(\nu + m' + 1)}\right]^{1/2} \frac{e^{i\pi(\nu+m')}}{(-m - m')!}$$

$$\times a^{-m-m'}(b^*)^{m'-m} F(-\nu - m, \nu - m + 1; -m - m' + 1; |a|^2)$$

$$(6)$$

For $m + m' > 0$,

$$C_{m',m}^{v}(M) = \left[\frac{\Gamma(v+m+1)\Gamma(v+m'+1)}{\Gamma(v-m+1)\Gamma(v-m'+1)}\right]^{1/2} \frac{e^{i\pi(v-m)}}{(m+m')!}$$

$$\times (a^*)^{m+m'} b^{v-m'} (b^*)^{v-m} F\left(-v+m, -v+m'; m+m'+1; -\left|\frac{a}{b}\right|^2\right)$$

$$= \left[\frac{\Gamma(v+m+1)\Gamma(v+m'+1)}{\Gamma(v-m+1)\Gamma(v-m'+1)}\right]^{1/2} \frac{e^{i\pi(v-m)}}{(m+m')!}$$

$$\times (a^*)^{m+m'} (b^*)^{m'-m} F(-v+m', v+m'+1; m+m'+1; |a|^2)$$

$$(7)$$

For the case v equals an integer or half-integer, the infinite matrices $B^v(M)$ and $C^v(M)$ coincide and give back the usual finite-dimensional matrices.

The existence of products of these matrices for particular values of the parameters a and b can be established by operating the product $\bar{M} = M'M$ on the spinor x. We then have

$$f_m^v[(M'M)^{-1}x] = \sum_{m''=-\infty}^{+\infty} B_{m'',m}^v(M'M) f_{m''}^v(x)$$

$$= f_m^v(M^{-1}x')$$

$$(8)$$

where

$$x' = (M')^{-1}x$$

that is,

$$u_1' = a'^*u_1 - b'u_2 \qquad u_2' = b'^*u_1 + a'u_2$$

$$(9)$$

But now we cannot set $|u_1'|$ and $|u_2'|$ equal to 1 arbitrarily; actually, $|u_1'|^2 = 1 - 2\text{Re}(a'b'u_1^*u_2)$ and $|u_2'|^2 = 1 + 2\text{Re}(a'b'u_1^*u_2)$. In this case, $f_m^v(M^{-1}x')$ can be expressed as a power series only if

$$\frac{|a|}{|b|} > \left[\frac{1+2|a'||b'|}{1-2|a'||b'|}\right]^{1/2}$$

By comparison, we find that

$$\sum_{m'=-\infty}^{+\infty} B_{m'',m'}^v(M') B_{m',m}^v(M) = B_{m'',m}^v(M'M)$$

$$(10)$$

at least if

$$|\bar{a}| > |\bar{b}| >; \qquad |a'| > |b'|$$

and

$$|a| > \left[\frac{1+2|a'||b'|}{1-2|a'||b'|}\right]^{1/2} |b|$$

$$(11)$$

These conditions can be made less restrictive by using the properties of $F(a_1, a_2; a_3; z)$ for a_1, a_2, z fixed and $a_3 \to \infty$.

A similar procedure cannot be adopted for the C-matrices since the conditions analogous to (11) in this case,

$$|a'| < |b'|$$

and

$$|a| < \left[\frac{1 - 2|a'||b'|}{1 + 2|a'||b'|}\right]^{1/2} |b|$$

are inconsistent with $|\bar{a}| < |\bar{b}|$.

But by re-expressing C in terms of B, we have

$$\sum_{m'=-\infty}^{+\infty} C_{m'',m'}^v([a', b'])C_{m',m}^v([a, b])$$

$$= e^{i\pi(2v+m''-m)} \times \sum_{-m'=-\infty}^{+\infty} B_{m'',-m'}^v([b', a'])B_{-m',m}^v([b, a])$$

$$= e^{i\pi(2v+m''-m)} B_{m'',m}^v(M'M)$$

for

$$\left|\frac{b'b}{a'a}\right| \geqslant 1 \qquad \frac{b'b^*}{a'a} \neq 1 \tag{12}$$

Similarly,

$$\sum_{m'=-\infty}^{+\infty} C_{m'',m'}^v(M')B_{m',m}^v(M) = C_{m'',m}^v(M'M) \tag{13}$$

for

$$\left|\frac{b'a}{a'b}\right| \geqslant 1 \qquad \frac{-b'a^*}{a'b} \neq 1$$

and

$$\sum_{m'=-\infty}^{+\infty} B_{m'',m'}^v(M')C_{m',m}^v(M) = C_{m'',m}^v(M'M) \tag{14}$$

for

$$\left|\frac{a'b}{b'a}\right| \geqslant 1 \qquad \frac{-a'b}{b'a^*} \neq 1$$

In particular, we have

$$B^v(M^{-1})B^v(M) = I \tag{15}$$

where $|a| \geqslant |b|$, and

$$C^v(M^{-1})C^v(M) = e^{i\pi(2v+m''-m)} \tag{16}$$

where $|a| < |b|$. I is the unit matrix of infinite order.

These and some further properties of the two sets of matrices B and C can be summarized as follows:

1. $M = [a, b]$ can be mapped into two sets of infinite-dimensional matrices $B^v(M), C^v(M)$; for $a = 0$ or $b = 0$, only one representative exists, the other becoming singular.
2. A product is defined for some pairs of matrices.
3. An identity exists.
4. Both left and right cancellations hold whenever the product is defined.
5. For every M in the unitary unimodular group, one of the corresponding infinite set of matrices has an inverse.
6. For a given sequence $M_1, M_2 \cdots M_n = \overline{M}$, the product of the corresponding infinite matrices with a fixed ordering rule exists for two different choices of representatives, leading to $B^v(\overline{M})$ and $C^v(\overline{M})$.
7. The associative law holds whenever all the products involved exist.

Structures with the above properties have been called "associative half-loops" (though loops, in general, are defined as objects which have all the properties of a group except associativity).

As is apparent, wherever the pair of matrices B and C is defined, it represents the elements of the representation. However, if one restricts attention to matrices in the neighborhood of identity, one obtains a "local representation" of the unitary unimodular group such as the one obtained by Andrews and Gunson, who have made further study of these representations and defined a Clebsch–Gordan coefficient for the addition of complex angular momenta. Using the common notation $D^v_{m',m}(\alpha, \beta, \gamma)$ for the B and C matrices wherever they are defined, and using the homomorphism of M with the elements of the rotation group R_3 by setting

$$a = \pm e^{-i(\alpha/2)} \left[\cos(\beta/2)\right] e^{-i(\gamma/2)}$$
$$b = \mp e^{-i(\alpha/2)} \left[\sin(\beta/2)\right] e^{i(\gamma/2)}$$

where α, β, and γ are the Euler angles, we can write

$$D^v_{m,m'}(\alpha, \beta, \gamma) = e^{im\alpha} d^v_{m,m'}(\cos\beta) e^{im'\gamma} \tag{17}$$

Andrews and Gunson also define functions of the "second kind" by

$$E_{m,m'}^{\nu}(\alpha, \beta, \gamma) = e^{im\alpha} e_{m,m'}^{\nu}(\cos \beta) e^{im'\gamma} \tag{18}$$

where

$$e_{m,m'}^{\nu}(z) = \frac{\pi}{2 \sin \pi(\nu - m)} \{ e^{\mp i\pi(\nu - m)} d_{m,m'}^{\nu}(z) - d_{m,-m'}^{\nu}(-z) \} \tag{19}$$

The signs $\mp$ are for Im $z \lessgtr 0$. For the e-functions, Andrews and Gunson establish the addition law

$$e_{m_1,m_1'}^{\nu_1}(z) e_{m_2,m_2'}^{\nu_2}(z) =$$

$$\sum_{\nu = \nu_1 + \nu_2 + 1}^{\infty} G(\nu_1, \nu_2; \nu; m_1, m_2) G(\nu_1, \nu_2; \nu; m_1', m_2') e_{m,m'}^{\nu}(z) \tag{20}$$

where

$$G(\nu_1, \nu_2; \nu; m_1, m_2) = \left[\frac{\pi \tan \pi(\nu - m)}{\tan \pi(\nu_1 - m_1) \tan \pi(\nu_2 - m_2)} \right]^{1/2}$$

$$\times C(-\nu_1 - 1, -\nu_2 - 1; -\nu - 1; -m_1, -m_2) \tag{21}$$

$$C(\nu_1, \nu_2; \nu; m_1, m_2) = \left[\frac{\Gamma(\nu_1 + m_1 + 1)\Gamma(\nu_2 - m_2 + 1)}{\Gamma(\nu_1 - m_1 + 1)\Gamma(\nu_2 + m_2 + 1)} \right.$$

$$\left. \times \Gamma(\nu - m + 1)\Gamma(\nu + m + 1) \right]^{1/2}$$

$$\times (2\nu + 1)^{1/2} \left[\frac{\Gamma(\nu_1 - \nu_2 + \nu + 1)\Gamma(-\nu_1 + \nu_2 + \nu + 1)}{\Gamma(\nu_1 + \nu_2 - \nu + 1)\Gamma(\nu_1 + \nu_2 + \nu + 2)} \right]^{1/2}$$

$$\times \frac{1}{\Gamma(\nu - \nu_2 + m_1 + 1)\Gamma(\nu - \nu_1 - m_2 + 1)}$$

$$\times {}_3F_2 \left[\begin{matrix} -\nu_1 + m_1, -\nu_2 - m_2, -\nu_1 - \nu_2 + \nu \\ \nu - \nu_1 - m_2 + 1, \nu - \nu_2 + m_1 + 1 \end{matrix} \right] \tag{22}$$

For integral values of $\nu_1 - m_1$ and $\nu_2 - m_2$, (22) reduces to the real Clebsch–Gordan coefficients.

We consider next the little group associated with a space-like four-vector which leaves invariant the form $p_t^2 - p_x^2 - p_y^2$. Bargmann has studied this group and has found both continuous and discrete representations for it. (The square of a space-like momentum vector has the meaning of the square of a momentum transfer.) The group contains the reflection T, which commutes with the generators of the group and has the eigenvalue τ. Assuming the symmetry of the integral equation for a scattering amplitude under this group, Sertorio and Toller find, on solving the equation by the Fredholm method, that the index τ disappears from the Fredholm determinant only if the direct potential is used, but remains if both a direct and an exchange potential are present. They interpret τ as the "signature" of the amplitude.

We can adopt a procedure similar to the one for the rotation group to obtain representations when the index ν labeling it is complex. Using again the homomorphism with two-dimensional matrices, we see that the matrix which leaves the third component of the four-vector x, written in the form

$$\begin{pmatrix} x_3 - x_4 & x_1 + ix_2 \\ x_1 - ix_2 & x_3 + x_4 \end{pmatrix} \tag{23}$$

invariant is the matrix

$$\begin{pmatrix} a & b \\ b^* & a^* \end{pmatrix} \tag{24}$$

with the unimodularity condition

$$aa^* - bb^* = 1$$

Equation (24) shows that the matrix is no longer unitary, but we can obtain the representations (whenever they exist) for this group from the corresponding ones for the rotation group by the replacement $-b^* \rightarrow b^*$. We get a clearer picture if we use the Euler angles for this case:

$$a = \cosh \zeta e^{-i(\mu+\nu)} \qquad b = \sinh \zeta e^{i(\nu-\mu)} \tag{25}$$

Here, ζ varies between 0 and ∞ and μ and ν between $-\pi$ and π. Since a can never be zero, we have for the matrices corresponding to B complete group properties without any singular point, while for the matrices C, $b = 0$ is still a singular point.

The spinor representations of the homogeneous Lorentz group were studied by Harishchandra, Gelfand, Bargmann, and others. The unitary representations are infinite-dimensional (apart from a trivial one-dimensional representation) In general, the representations are characterized by two parameters l_0 and l_1, the first referring to the lowest weight of the corresponding rotation subgroup participating in the representation, and l_1 is any arbitrary complex number:

$$l_0 = j - j' \qquad l_1 = j + j' + 1 \tag{26}$$

$2j$ and $2j'$ are the number of undotted and dotted indices in a general spinor of rank $2(j + j')$ defined in a $(2j + 1)(2j' + 1)$ dimensional space. (Since the group to which the homogeneous Lorentz group is homomorphic is the unimodular group which is not unitary, the complex-conjugate spinor transforms in a different way from the spinor, hence the two types of indices for the general spinor). $j + j'$ gives the highest spin value corresponding to the representation, and one notices from (26) that it can be complex. For the infinite-dimensional unitary representations, in particular it is given by $-1 + ik$, where k is real.

We can study the representations for complex spin (i. e., complex ν and ν') as in the previous cases. The general spinor is a monomial of the type

$$f_{m,m'}^{\nu\nu'}(x, \dot{x}) = A_{mm'}^{\nu\nu'} u_1^{\nu-m} u_2^{\nu+m} u_1^{\nu'-m'} u_2^{\nu'+m'} \tag{27}$$

The undotted spinor $\chi = \begin{pmatrix} u_1 \\ u_2 \end{pmatrix}$ is acted on by the unimodular matrix $M = \begin{pmatrix} ab \\ cd \end{pmatrix}$, and the dotted spinor $\dot{x}$ by the matrix $M^* = \begin{pmatrix} a^*b^* \\ c^*d^* \end{pmatrix}$. Four cases are possible:

1. $|d| \geqslant |b|$; $|a| \geqslant |c|$; $d \neq 0$, $a \neq 0$
For $m'' - m \leqslant 0$; $m''' - m' \leqslant 0$,

$$B_{m''m,m'''m'}^{\nu\nu'}(M, M^*) = \frac{A_{mm'}^{\nu\nu'}(-1)^{m-m''+m'-m'''}}{A_{m''m'''}^{\nu\nu'}(m - m'')!(m' - m''')!}$$

$$\times \frac{\Gamma(\nu + m + 1)\Gamma(\nu' + m' + 1)}{\Gamma(\nu + m'' + 1)\Gamma(\nu' + m''' + 1)}$$

$$\times a^{\nu+m''}(a^*)^{\nu'+m'''} c^{m-m'}(c^*)^{m'-m'''}$$

$$\times d^{\nu-m}(d^*)^{\nu'-m'} F\left(-\nu + m, -\nu - m''; m - m'' + 1; \frac{bc}{ad}\right)$$

$$\times F\left(-\nu' + m', -\nu' - m'''; m' - m''' + 1; \frac{b^*c^*}{a^*d^*}\right)$$

For $m''' - m \leqslant 0$; $m''' - m' > 0$, make the changes $c^* \to b^*$; $m' \longleftrightarrow m'''$.

For $m'' - m > 0$; $m''' - m' \leqslant 0$, make the changes $c \to b$; $m \longleftrightarrow m''$.

For $m'' - m > 0$; $m''' - m' > 0$, make the changes $c \to b$, $c^* \to b^*$; $m \longleftrightarrow m''$, $m' \longleftrightarrow m'''$. (28)

2. $|d| \leqslant |b|$, $|a| \leqslant |c|$; $b \neq 0$, $c \neq 0$
For $m + m'' \leqslant 0$; $m' + m''' \leqslant 0$,

$$C^{vv'}_{m''m,m'''m'}(M, M^*) = \frac{A^{vv'}_{mm'}(-1)^{m+m''+m'+m'''}}{A^{vv'}_{m''m'''}(-m-m'')!(-m'-m''')!}$$

$$\times \; e^{2i\pi(v+v')}\frac{\Gamma(v+m+1)\Gamma(v'+m'+1)}{\Gamma(v-m''+1)\Gamma(v'-m'''+1)}$$

$$\times \; a^{m+m''}(a^*)^{m'+m''}b^{v-m}(b^*)^{v'-m'}c^{v-m''}(c^*)^{v'-m'''}$$

$$\times \; F\left(-v+m, -v+m''; -m-m''+1; \frac{ad}{bc}\right)$$

$$\times \; F\left(-v'+m'; -v'+m'''; -m'-m'''+1; \frac{a^*d^*}{b^*c^*}\right)$$

For $m+m'' \leqslant 0$; $m'+m''' > 0$, make the changes $a^* \to d^*$; $m' \longleftrightarrow -m'''$.

For $m+m'' > 0$; $m'+m''' \leqslant 0$, make the changes $a \to d$, $m \longleftrightarrow -m'$.

For $m+m'' > 0$; $m'+m''' > 0$, make the changes $a \to d$, $a^* \longleftrightarrow d^*$, $m \longleftrightarrow -m''$, $m' \longleftrightarrow -m'''$. (29)

In the above

$$A^{vv'}_{mm'} = \frac{\Gamma(2v+1)\Gamma(2v'+1)}{\Gamma(v+m+1)\Gamma(v-m+1)\Gamma(v'+m'+1)\Gamma(v'-m'+1)}$$

3. $|d| \geqslant |b|$; $|a| \leqslant |c|$

and

4. $|d| \leqslant |b|$; $|a| \geqslant |c|$

In both these cases, v has to be an integer or a half-integer so that they correspond to the usual finite-dimensional (non-unitary) representations of the homogeneous Lorentz group.

For the 12-parameter complex Lorentz group, the dotted spinors do not transform according to the matrix M^* but another unimodular matrix N with parameters $\alpha, \beta, \gamma,$ and δ. There are sixteen possibilities of which twelve lead to infinite-dimensional matrix representation with singular points, the four others being finite-dimensional. In only four of the former cases, both v and v' can be complex. In four of the rest, one of the indices v or v' can take only integral or half-integral values, while for the remaining four it is the other way. Thus the algebra of matrices corresponding to the representations of this group presents an intricate pattern.

Finally, we shall briefly mention the consideration of Lurçat in which he criticizes the usual assumption that spin is an inessential complication in quantum field theory and plays no dynamical role. He defines wave functions and fields, not on Minkowski space but on the Poincaré manifold. He finds that in reconstructing the subspace of the Hilbert space from the two-point function (using the positivity condition on the scalar product and the spectral condition), the latter plays the role of a trace on a group in the construction of a representation of a group. The two-point function of a free-field with mass m and spin s is the character of the irreducible representation $[m, s]$. For interacting particles, a theorem on traces on a group is used to give a spectral representation for the Green's function in which the generalized spectral measure $\rho(m^2, s)$ depends not only on mass but also on spin. For each value of m^2, it defines a spin spectrum which is a generalization of the notion of the Regge trajectory.

This Poincare field theory is supposed to be in furtherance of the program defined by Dirac for finding "representations of the inhomogeneous Lorentz group corresponding to physical reality." The usual representations, as given by Wigner, correspond to noninteracting particles. Assuming the momentum and angular momentum generators to be kinematic generators, and the energy and pure Lorentz transformation operators to be dynamic generators, Dirac derives conditions which a quantum field theory that is relativistic has to obey. For the case of spins 0, $\frac{1}{2}$, and 1, the condition is

$$[T_{00}(x), T_{00}(x')] = -i(T_{0k}(x) + T_{0k}(x'))\partial_k \delta(\mathbf{x} - \mathbf{x}') \qquad (31)$$

where $x_0 = x_0'$ and T_{00} and T_{0k} are the energy and momentum densities, respectively. This result was also arrived at independently by Schwinger.

REFERENCES

1. T. Regge, *Nuovo Cimento* **14**: 951 (1959); **18**: 947 (1960).
2. D. Amati, A. Stanghellini, and S. Fubini, *Nuovo Cimento* **26**: 896 (1962).
3. E. G. Beltrametti and G. Luzzatto, *Nuovo Cimento* **29**: 1003 (1963).
4. M. Andrews and J. Gunson, *J. Math. Phys.* **5**: 1391 (1964).
5. V. Bargmann, *Ann. Math.* **48**: 568 (1947).
6. L. Sertorio and M. Toller, *Nuovo Cimento* **33**: 413 (1964).
7. Harish-Chandra, *Proc. Roy. Soc. A* **183**: 284 (1945).
8. I. M. Gelfand, R. A. Minlos, and Z. Ya. Shapiro, *Representations of the Rotation and Lorentz Groups and Their Applications*, Pergamon Press, Oxford, 1963.
9. K. Venkatesan (unpublished).
10. F. Lurçat, Preprint.
11. E. P. Wigner, *Ann. Math.* **40**: 149 (1939).
12. P. A. M. Dirac, "La théorie quantique des champs" (Solvay meeting 1961), New York, 1962. *Rev. Mod. Phys.* **34**: 592 (1962).
13. J. Schwinger, *Phys. Rev.* **130**: 406, 800 (1963).

Muon Capture by Complex Nuclei

V. DEVANATHAN

UNIVERSITY OF MADRAS
Madras, India

The direct experimental study of the basic elementary process $\mu^- + p \rightarrow n + \nu$ is not possible, and so one has to extract the information on this process from the measurement of muon capture rate by complex nuclei or from a study of the muon capture in liquid hydrogen. The complications that arise out of complexities of nuclear and molecular physics are therefore unavoidable.

Below is outlined the theory of muon capture by complex nuclei. The problem is twofold. One part is the study of the precise form of the muon capture interaction and the determination of the coupling constants, and the other is concerned with the details of nuclear structure. The interaction Hamiltonian of the basic elementary process itself is not yet well known. Although the universal Fermi interaction is assumed to be operative in the weak processes such as β-decay, μ-decay, and μ-capture, only μ-decay is free from the effects of strong interactions. The structure of the interaction Hamiltonian for μ-capture is considerably affected by the presence of strong interaction currents. The fundamental universal "bare" four-fermion interaction, which is assumed to be a vector-axial interaction, is modified to yield other types of interaction, namely, pseudoscalar, scalar, tensor, and weak magnetism. The determination of these coupling constants is of main interest. This may be done by choosing nuclei with closed shells, the structure of which presents very little uncertainty. Once the coupling constants are determined, muon capture may be used as a tool to probe the nuclear structure.

In view of these vast potentialities, the problem of muon capture by complex nuclei is here treated in most general terms without restricting the discussion to any particular nucleus. Starting with the effective Hamiltonian of Fujii and Primakoff,[1] the capture rate is deduced using the shell model for the nucleus. This involves the summing and averaging over the leptonic and nuclear spin states. The usual procedure[2] is first to average over initial muon states and sum over final neutrino states and then to evaluate the matrix element between the nuclear states. It is found that the reversal of this operation results in great simplicity. Then the problem is exactly identical with the one which has been treated elsewhere.[3]

The capture rate depends on the spatial wave function of the nucleon in the nucleus and also the type of coupling (j-j or L-S or intermediate coupling) that is assumed to be operative between the nucleons, and hence the capture rate alone will not yield reliable and conclusive results regarding the basic interaction process which is responsible for muon capture. Therefore, it is suggested that a study of the polarization of the outgoing nucleon in the process $\mu^- + X \rightarrow Y + N + \nu$ is desirable since it is unaffected by the nuclear wave function.

1. THE EFFECTIVE HAMILTONIAN

The complete matrix element (which includes the effects of strong interactions of the nucleons) for the process of muon capture can be written as

$$
\begin{aligned}
M = \frac{1}{\sqrt{2}} \{ & (\bar{u}_\nu (1 - \gamma_5) i \gamma_\lambda \gamma_5 u_\mu) \\
\times & [A(\bar{u}_n i \gamma_\lambda \gamma_5 u_p) - B(\bar{u}_n q_\lambda \gamma_5 u_p) + E(\bar{u}_n \sigma_{\lambda\rho} q_\rho \gamma_5 u_p)] \\
+ & (\bar{u}_\nu (1 - \gamma_5) \gamma_\lambda u_\mu) [C(\bar{u}_n \gamma_\lambda u_p) \\
& - iD(\bar{u}_n \sigma_{\lambda\rho} q_\rho u_p) + iF(\bar{u}_n q_\lambda u_p)] \}
\end{aligned}
\tag{1}
$$

The quantities A, B, C, D, E, and F are form factors which depend on the invariant four-momentum transfer q^2, and they are nearly constant if q^2 does not vary too much. Also, they are real if T-invariance holds.

Weinberg[4] distinguishes between interactions of the first class (the terms with coefficients A, B, C, and D) and those of the second class (the terms with coefficients E and F). Goldberger and Treiman[5] and Fujii and Primakoff[1] assume explicitly that there are no second-class interactions; however, this should be established experimentally.

A, B, C, D, E, and F may be considered as kinds of apparent coupling constants. In the following notation, all coupling constants have the usual dimension of the four-fermion coupling constant:

$$C = g_V \quad \text{vector apparent coupling constant}$$
$$A = g_A \quad \text{axial vector apparent coupling constant}$$
$$m_\mu B = g_P \quad \text{(induced) pseudoscalar coupling constant}$$
$$2MD = g_M \quad \text{weak-magnetism coupling constant}$$
$$m_\mu F = g_S \quad \text{(induced) scalar coupling constant}$$
$$2ME = g_T \quad \text{(induced) tensor coupling constant}$$

where m_μ is the mass of the muon and M is the nucleon mass.

It is assumed that the fundamental universal "bare" four-fermion interaction is a vector-axial interaction with coupling constants $g_V^0 = -g_A^0$. The strong interactions of the nucleons cause renormalizations of the nucleon currents such that additional terms appear and also alter the values of A and C.

Using the four-momentum conservation law,

$$p_\lambda - n_\lambda = \nu_\lambda - \mu_\lambda$$

and the free Dirac equations for u_ν, u_μ, u_p, and u_n,

$$(\gamma_\lambda p_\lambda - im)u = 0$$

and also the relation

$$p_\lambda p_\lambda = -m^2$$

we obtain

$$M = \frac{1}{\sqrt{2}}\Big\{A[\bar{u}_\nu(1 - \gamma_5)i\gamma_\lambda\gamma_5 u_\mu](\bar{u}_n i\gamma_\lambda\gamma_5 u_p)$$
$$+ m_\mu B[\bar{u}_\nu(1 - \gamma_5)\gamma_5 u_\mu](\bar{u}_n\gamma_5 u_p)$$
$$- \frac{iC}{M}[\bar{u}_\nu(1 - \gamma_5)\gamma_\lambda u_\mu](\bar{u}_n p_\lambda u_p)$$
$$- \frac{iC}{2M}[\bar{u}_\nu(1 - \gamma_5)\gamma_\lambda(\mu_\lambda - \nu_\lambda)u_\mu](\bar{u}_n u_p)$$
$$+ i\Big(\frac{C}{2M} + D\Big)[\bar{u}_\nu(1 - \gamma_5)\gamma_\lambda(\mu_\rho - \nu_\rho)u_\mu](\bar{u}_n\sigma_{\lambda\rho} u_p)\Big\}$$

The terms involving E and F have been neglected in the above.

The transition matrix element M calculated in a nonrelativistic approximation for the muons and nucleons corresponds to a suitably chosen nonrelativistic effective Hamiltonian $\mathcal{H}_{\text{eff}}$. The $\mathcal{H}_{\text{eff}}$ generalized to a many-nucleon problem is related to the corresponding matrix element by

$$M = \langle \text{final state of } A \text{ nucleons } | \mathcal{H}_{\text{eff}} | \text{ initial state of } A \text{ nucleons} \rangle \quad (3)$$

and describes muon capture by an aggregate of A nucleons. Retaining only the $1/M$ term and neglecting terms of higher order, $\mathcal{H}_{\text{eff}}$ can be written as

$$\mathcal{H}_{\text{eff}} = \frac{1}{\sqrt{2}} \bar{u}_\nu \tau_L^{(+)} \frac{1 - \boldsymbol{\sigma}_L \cdot \hat{v}}{\sqrt{2}} \sum_{i=1}^{A} \tau_i^{(-)} [G_V \mathbf{l}_i \cdot \mathbf{l}_i + G_A \boldsymbol{\sigma}_L \cdot \boldsymbol{\sigma}_i$$

$$- G_P(\boldsymbol{\sigma}_L \cdot \hat{v})(\boldsymbol{\sigma}_i \cdot \hat{v}) - \frac{g_V}{MC}(\boldsymbol{\sigma}_L \cdot \hat{v})(\boldsymbol{\sigma}_L \cdot \mathbf{p}_i)$$

$$- \frac{g_A}{MC}(\boldsymbol{\sigma}_L \cdot \hat{v})(\boldsymbol{\sigma}_i \cdot \mathbf{p}_i)] u_\mu \quad (4)$$

This may be written in a more compact form:

$$\mathcal{H}_{\text{eff}} = \sum_{i=1}^{A} (\boldsymbol{\sigma}_i \cdot \mathbf{K}_i + \mathbf{L}_i) e^{-i\boldsymbol{v} \cdot \mathbf{r}_i} \varphi_\mu(\mathbf{r}_i) \tau_i^{(-)} \quad (5)$$

where

$$\mathbf{K}_i = \tfrac{1}{2}(1 - \boldsymbol{\sigma}_L \cdot \hat{v})\left\{ G_A \boldsymbol{\sigma}_L - G_P(\boldsymbol{\sigma}_L \cdot \hat{v})\hat{v} - \frac{g_A}{MC}(\boldsymbol{\sigma}_L \cdot \hat{v})\mathbf{p}_i \right\} \quad (6)$$

$$\mathbf{L}_i = \tfrac{1}{2}(1 - \boldsymbol{\sigma}_L \hat{v})\left\{ G_V \mathbf{l}_L \mathbf{l}_i - \frac{g_V}{MC}(\boldsymbol{\sigma}_L \cdot \hat{v})(\boldsymbol{\sigma}_L \cdot \mathbf{p}_i) \right\} \quad (7)$$

In the above, $\tau_i^{(-)}$, $\mathbf{l}_i$, $\boldsymbol{\sigma}_i$, $\mathbf{p}_i$ are operators for the nucleon; $\boldsymbol{\sigma}_L$ and $\mathbf{l}_L$ are the lepton operators; and $\hat{v}$ is the unit vector in the direction of the neutrino momentum $\boldsymbol{v}$. G_V, G_A, and G_P might be called effective coupling constants, and they are given by

$$G_V = C\left(1 + \frac{v}{2M}\right) = g_V^{(\mu)}\left(1 + \frac{v}{2M}\right)$$

$$G_A = A - \left(\frac{C}{2M} + D\right)v = g_A^{(\mu)} - (g_V^{(\mu)} + g_M^{(\mu)})\frac{v}{2M}$$

$$G_P = (m_\mu B - A)\frac{v}{2M} - \left(\frac{C}{2M} + D\right)v$$

$$= [(g_P^{(\mu)} - g_A^{(\mu)}) - (g_V^{(\mu)} + g_M^{(\mu)})]\frac{v}{2M} \quad (8)$$

2. PARTIAL MUON CAPTURE RATE

The muon capture rate for a transition from an initial nuclear state $|i\rangle$ to a final nuclear state $|f\rangle$ is given by (suppressing the isotopic spin quantum numbers)

$$\Lambda_{if} = \frac{\nu^2}{2\pi} \frac{1}{2J_i + 1} \sum_{M_i} \sum_{M_f} |\langle J_f M_f| \sum_{i=1}^{A} (\boldsymbol{\sigma}_i \cdot \mathbf{K}_i + \mathbf{L}_i)e^{-i\nu\cdot\mathbf{r}_i} \varphi_\mu(\mathbf{r}_i)|J_i M_i\rangle|^2$$

$$(9)$$

The quantities $\mathbf{K}_i$ and $\mathbf{L}_i$ involve lepton spin operators which are evaluated between the lepton spin states at the end. The usual procedure adopted by Luyton and colleagues[2] is first to sum and average over the lepton spins and then to consider the nuclear problem. In the present treatment, the sequence of operations is reversed. The reversal of the operation yields a form to the structure of the transition operator which we are already familiar with and whose evaluation between any two nuclear states has been discussed.

As a first step, the muon wave function may be taken as a constant over nuclear volume:

$$\Lambda_{if} = \frac{\nu^2}{2\pi} |\varphi_\mu|^2 \frac{1}{2J_i + 1} \sum_{M_i} \sum_{M_f} |\langle J_f M_f|T|J_i M_i\rangle|^2 \qquad (10)$$

where

$$T = \sum_i t_i = \sum_i (\boldsymbol{\sigma}_i \cdot \mathbf{K}_i + \mathbf{L}_i)e^{-i\nu\cdot\mathbf{r}_i}$$

$$= \sum_{\substack{i \\ n=0,1}} (\boldsymbol{\sigma}_{i,n} \cdot \mathbf{K}_{i,n})e^{-i\nu\cdot\mathbf{r}_i} \qquad (11)$$

Expanding $e^{-i\nu\cdot\mathbf{r}_i}$ in terms of spherical harmonics, the transition operator can be written as a sum of spherical tensor operators

$$t_i = \sum_\lambda t_i^{\lambda,m_\lambda}$$

Using the Wigner-Eckart theorem, the dependence over the magnetic quantum numbers may be separated out, and the reduced matrix element can be further simplified using the shell model description for the nucleus. If there are n equivalent nucleons outside the closed shells, the transition matrix element can be written as

$$\langle J_f\|\sum_i t_i\|J_i\rangle = n\langle J_f\|t_i\|J_i\rangle$$

$$= n \times \text{(matrix element between single nucleon states)}$$

$$\times \text{(parentage overlap)} \qquad (12)$$

Thus, the many-nucleon problem is reduced to the problem of evaluation of the single-nucleon matrix element. The evaluation of the single nucleon matrix element is straightforward and is outlined in reference 3. Here we only state the results. The single-nucleon transition probability obtained after summing over the final spin states and averaging over the initial spin states is

$$\sum_{m_i} \sum_{m_f} |\langle f_{n_f l_f}(r); l_f \tfrac{1}{2} J_f m_f |(\boldsymbol{\sigma}\cdot\mathbf{K} + L)e^{-i\boldsymbol{\nu}\cdot\mathbf{r}}| l_i \tfrac{1}{2} j_i m_i; f_{n_i} l_i(r)\rangle|^2$$

$$= \sum_{l,n} \sum_{l',n'} \sum_{\lambda,m_\lambda} 4\pi(i)^{l-l'} F_l F_{l'}^*(-1)^{l+l'+n+n'}$$

$$[Y_l(\hat{\nu}) \times K_n]_\lambda^{-m_\lambda} [Y_{l'}(\hat{\nu}) \times K_{n'}]_\lambda^{-m_\lambda *}$$

$$[J_f]^2[J_i]^2[l_i]^2[\tfrac{1}{2}]^2[l][l'][n][n']$$

$$C(l_i l l_f; 00)C(l_i l' l_f; 00)$$

$$\begin{Bmatrix} l_i & l & l_f \\ 1/2 & n & 1/2 \\ J_i & \lambda & J_f \end{Bmatrix} \begin{Bmatrix} l_i & l' & l_f \\ 1/2 & n' & 1/2 \\ J_i & \lambda & J_f \end{Bmatrix}$$

where
$$[J] = (2J + 1)^{1/2}$$

and
$$F_l = \int_0^\infty f_{n_f l_f}^* j_l(\nu r) f_{n_i l_i}(r) r^2 \, dr$$

In (13), n and n' can take only two values, 0 and 1, and the values of l and l' are highly restricted by the parity CG coefficients. The Wigner nine j-symbols also restrict the value of λ, and thus the calcuation can be performed with ease. Further,

$$\sum_{m_\lambda} [Y_l(\hat{\nu})\times K_n]_\lambda^{-m_\lambda} [Y_l(\hat{\nu})\times K_n^*]_\lambda^{-m_\lambda *}$$

$$= \sum_N \frac{(-1)^{\lambda+N}}{(4\pi)^{1/2}}[\lambda]^3[l][l']C(ll'N; 00)\begin{Bmatrix} l & n & \lambda \\ l' & n' & \lambda \\ L & N & 0 \end{Bmatrix}[Y_N(\hat{\nu})]\cdot K_N \quad (14)$$

In the above summation, N can take values either 0 or 2, and the scalar product $[Y_N(\hat{\nu})\cdot K_N]$ is evaluated and presented in Table I.

Table   I

N	n	n'	$[Y_N(V)\cdot K_N]$
0	0	0	$[4\pi]^{-1/2} LL*$
0	1	1	$-(12\pi)^{-1/2}\mathbf{K}\cdot\mathbf{K}*$
2	1	1	$-\frac{1}{2K^2}\left(\frac{15}{2\pi}\right)^{1/2}\{(\boldsymbol{\nu}\cdot\mathbf{K})(\boldsymbol{\nu}\cdot\mathbf{K}*)-\frac{1}{3}\nu^2(\mathbf{K}\cdot\mathbf{K}*)\}$

The procedure outlined above is not only rigorous but more general and applicable to any nucleus. As a first approximation, the velocity-dependent terms in L and K can be neglected. It is shown that these terms contribute less than 10% to the capture rate.

3. POLARIZATION OF THE OUTGOING NUCLEON IN THE MUON CAPTURE PROCESS

The determination of the partial capture rate depends on (a) the potential well of the nucleus, and through that the spatial wave function of the nucleon in the nucleus; and (b) the type of coupling $(j - j, L - S,$ or intermediate coupling) that exists between the nucleons. Hence, the partial capture rate alone will not yield reliable results on the structure of the interaction Hamiltonian responsible for the elementary muon capture process. So, one has to look for other processes which may yield results of greater reliability. It is suggested that the polarization of the outgoing nucleon in the capture process

$$\mu' + x \rightarrow Y + N + \nu$$

is uninfluenced by the aforementioned two factors, and hence an experimental study of this is recommended for the unique determination of the type and magnitude of the muon capture coupling constants.

The amplitude for the muon capture by a bound nucleon, with the nucleon going out finally as a free particle, can be written as

$$\langle \mathbf{p}, \tfrac{1}{2}m_s | (\boldsymbol{\sigma} \cdot \mathbf{K} + L)e^{-i\nu \cdot \mathbf{r}} \varphi_i(\mathbf{r}) | u(r); l\tfrac{1}{2}jm \rangle \tag{15}$$

where $\mathbf{p}$ is the momentum of the outgoing nucleon, $u(r)$ is its radial wave function when bound, l is the initial orbital, $\tfrac{1}{2}$ is the spin, j is the total angular momentum, and m is its projection quantum number.

If there are initially n equivalent nucleons outside closed shells forming a total nuclear spin J_i, and J_f denotes the spin of the residual nucleus, then the amplitude for the nuclear process is simply

$$n \langle (j)^{n-1} J_f T_f; j\tau | \} (j)^n J_i T_i \rangle C(J_f j J_i; M_f m M_i)$$

$$C(T_f \tfrac{1}{2} T_i; M_{T_f}, M_{T_i} - M_{T_f}) \langle \tfrac{1}{2} m_s | (\boldsymbol{\sigma} \cdot \mathbf{K} + L) e^{-i(\nu + \mathbf{p}) \cdot \mathbf{r}} \varphi_i(\mathbf{r}) | u(r) l\tfrac{1}{2} jm \rangle \tag{16}$$

where $\langle (j)^{n-1} J_f T_f; j\tau |\} (j)^n J_i T_i \rangle$ denotes the coefficient of fractional parentage; T_f, T_i, M_{T_f}, and M_{T_i} are the nuclear isotopic spin quantum numbers; and M_f and M_i the nuclear magnetic quantum numbers. Now the density matrix representing polarization of the outgoing nucleon can be written as

$$\rho_{m_s m_{s'}} = n^2 |\langle (j)^{n-1} J_f T_f, j\tau |\} (j)^n J_i T_i \rangle|^2 \sum_{M_f, M_i} |C(J_f j J_i, M_f m M_i)|^2$$

$$\times |C(T_f \tfrac{1}{2} T_i; M_{T_f} - M_{T_i}, M_{T_i})|^2$$

$$\times \langle \tfrac{1}{2} m_s |(\boldsymbol{\sigma} \cdot \mathbf{K} + L) e^{-i(\nu + \mathbf{p}) \cdot \mathbf{r}} \varphi_\mu(r) | u(r), l\tfrac{1}{2} jm \rangle$$

$$\times \frac{1}{2J_i + 1} \langle \tfrac{1}{2} m_{s'} | e^{-i(\nu + \mathbf{p}) \cdot \mathbf{r}} (\boldsymbol{\sigma} \cdot \mathbf{K} + L) | u(r) l\tfrac{1}{2} jm \rangle^*$$

$$= n^2 |C(T_f \tfrac{1}{2} T_i; M_{T_f}, M_{T_i} - M_{T_i})|^2$$

$$\times |\langle (j)^{n-1} J_f T_f; j\tau |\} (j)^n J_i T_i \rangle|^2 \frac{2j + 1}{2J_i + 1} \rho_{m_s, m_{s'}}^{s.b.n.} \qquad (17)$$

where

$$\rho_{m_s, m_{s'}}^{s.b.n.} = \sum_m \langle \tfrac{1}{2} m_s |(\boldsymbol{\sigma} \cdot \mathbf{K} + L) e^{-i(\nu + \mathbf{p}) \cdot \mathbf{r}} \varphi_\mu(r) | u(r), l\tfrac{1}{2} jm \rangle$$

$$\times \frac{1}{2j + 1} \langle \tfrac{1}{2} m_{s'} | e^{-i(\nu + \mathbf{p}) \cdot \mathbf{r}} (\boldsymbol{\sigma} \cdot \mathbf{K} + L) \varphi_\mu(r) | u(r), l\tfrac{1}{2} jm \rangle^* \qquad (18)$$

denotes the density matrix of the outgoing free nucleon in the case of a single bound nucleon ($n = 1$) in the initial state. The density matrix (18) can be re-expressed in the form

$$\rho_{m_s, m_{s'}}^{s.b.n.} = \sum_\mu \sum_{\mu'} \langle \tfrac{1}{2} m_s |(\boldsymbol{\sigma} \cdot \mathbf{K} + L)| \tfrac{1}{2} \mu \rangle \qquad \rho_{\mu, \mu'}^{eff} \langle \tfrac{1}{2} m_{s'} |(\boldsymbol{\sigma} \cdot \mathbf{K} + L)| \tfrac{1}{2} \mu' \rangle^*$$

$$(19)$$

where

$$\rho_{\mu, \mu'}^{eff} = \frac{16\pi^2 |f_l|^2}{2j + 1} \sum_m C(l\tfrac{1}{2} j; m_l \mu m) C(l\tfrac{1}{2} j; m_{l'} \mu' m)$$

$$\times Y_{l, -m_l}(\nu + \mathbf{p}) Y_{l, -m_{l'}}(\nu + \mathbf{p}) \qquad (20)$$

in which f_l denotes the overlap integral

$$f_l = \int j_l(|\nu + \mathbf{p}| \cdot \mathbf{r}) u(r) r^2 \, dr \qquad (21)$$

One might anticipate that since the fictitious effective density matrix ρ^{eff} contains no spin effects, the corresponding initial nucleon polarization is zero. In fact, expression (20) can be simplified to yield

$$\rho_{\mu,\mu'}^{\text{eff}} = 2\pi |f_l|^2 \delta_{\mu,\mu'} \tag{22}$$

which thus forms an overall multiplying factor in $\rho_{m_n,m_s'}$ together with the factors appearing in (17). These multiplying factors cancel when we consider the observable polarization parameters **p** of the outgoing nucleon, which are therefore simply given by

$$\mathbf{p} = \frac{i[\mathbf{K}^* \times \mathbf{K} + L^*\mathbf{K} - L\mathbf{K}^*]}{\mathbf{K}^*\mathbf{K} + L^*L} \tag{23}$$

The above expression for the final nucleon polarization is obtained using a $j - j$ coupling scheme. The same result holds true even if the n equivalent nucleons outside closed shells obey $L - S$ or intermediate coupling schemes, as can be verified by explicit calculation following the same procedure.

REFERENCES

1. A. Fujii and H. Primakoff, *Nuovo Cimento* **12**: 327 (1959).
2. J. R. Luyton, H. P. C. Rood, and H. A. Tolhock *Nucl. Phys.* **41**: 236 (1963).
3. V. Devanathan and G. Ramachandran, *Nucl. Phys.* **42**: 254 (1963); G. Ramachandran and V. Devanathan, *Nucl. Phys.* (in press).
4. S. Weinberg, *Phys. Rev.* **112**: 1375 (1959).
5. M. L. Goldberger and S. B. Treiman, *Phys. Rev.* **111**: 354 (1958).

Comments on Sum Rules

P. T. LANDSBERG[†] AND D. J. MORGAN[†]

UNIVERSITY COLLEGE
Cardiff, Wales

1. INTRODUCTION

The purpose of this discussion is to give a reasonable, straightforward method of obtaining sum rules for any system, given the Hamiltonian of the system and the main variables which one wants to occur in the rules. A simple exposition of this work is given here. The detailed account has been published elsewhere.[‡]

Sum rules are important because they can be formulated for many-body problems, and are then normally even exact for the model considered. They represent relations for certain matrix elements. Thus, if the wave functions are known only approximately, sum rules can serve as a check on the approximation made. In other cases, sum rules have been used to gain preliminary ideas about properties of matrix elements which are initially unknown.

A few well-known results of quantum mechanics are recalled first (Section 2), and next they are shown to result from a simple unifying theory (Sections 3 and 4). The use of the theory to obtain new results is briefly indicated at the end.

2. MOTIVATION

The quantum states of a single- or many-particle system will be described by two quantum numbers n and l. The first identifies

[†]Department of Applied Mathematics and Mathematical Physics.
[‡]*Proc. Phys. Soc.* **86**: 261 (1965).

the energy level E_n, the second allows for degeneracy if n is given. We now summarize the following results:

A. A nonrelativistic particle has a homogeneous potential energy function of degree s. Its kinetic energy is T, its potential energy is V. The virial theorem then states[†]

$$2\langle nl|T|nl\rangle = s\langle nl|V|nl\rangle = \frac{2s}{2+s}E_n \tag{1}$$

For a Coulomb potential $s = -1$, and for a harmonic oscillator $s = 2$.

B. A Hamiltonian H and its eigenvalues E_n depend on a parameter α. Feynman's theorem[2] states that

$$\left\langle nl\left|\frac{\partial H}{\partial \alpha}\right|nl\right\rangle = \frac{\partial E_n}{\partial \alpha} \tag{2}$$

In particular, if $H = H_0 + \alpha V$, one has another expression

$$\langle nl|V|nl\rangle = \frac{\partial E_n}{\partial \alpha} \tag{3}$$

for the expectation value of the potential energy.

C. The original sum rule is associated with the names Thomas, Reiche, and Kuhn. It was used for N-electron atoms or ions and states:[3]

$$N + \frac{2}{m}\sum_{\substack{n'(\neq n)\\l'}}\frac{\left\langle nl\left|\sum_{w=1}^{N}p_{wi}\right|n'l'\right\rangle\left\langle n'l'\left|\sum_{w=1}^{N}p_{wi}\right|nl\right\rangle}{E_n - E_{n'}} = 0 \tag{4}$$

where p_{w_i} is the operator for the ith component of the linear momentum of electron w. It assumes a Hamiltonian of the form

$$\sum_{w=1}^{N}\frac{\mathbf{p}_w^2}{2m} + V(\mathbf{r}_1, \ldots, \mathbf{r}_N) \tag{4'}$$

that is, the absence of velocity-dependent forces.

D. For one electron in a periodic lattice,[4]

$$\delta_{ij} + \boxed{2}\,m\sum_{\substack{n''(\neq n)\\l''}}\frac{\langle nl'|v_i|n''l''\rangle\langle n''l''|v_j|nl\rangle}{E_n(\mathbf{k}) - E_{n''}(\mathbf{k})} = \left(\frac{1}{m^*}\right)_{ij}^{n,\mathbf{k}}m \tag{5}$$

Here, m is the normal electron mass, and the term on the right is the (i, j)-component of the reciprocal effective mass tensor for band n at wave vector $\mathbf{k}$. The velocity operator $\mathbf{v}$ may in the absence of

---

[†]See, for instance, Bethe and Salpeter.[1]

spin-orbit interaction be interpreted as $\mathbf{p}/m$. The [2] indicates that for every term written down behind it, there are really two different terms, the additional term having i and j interchanged.

E. If

$$\rho_k \equiv \sum_{j=1}^{N} \exp(i\mathbf{k}\cdot\mathbf{r}_j)$$

is the density fluctuation operator for a system having the Hamiltonian given under paragraph C, then its matrix elements satisfy

$$N + \frac{m}{\hbar^2 k^2} \sum_{n'l'} (E_n - E_{n'})[\langle nl|\rho_k^+|n'l'\rangle\langle n'l'|\rho_k|nl\rangle$$
$$+ \langle nl|\rho_k|n'l'\rangle\langle n'l'|\rho_k^+|nl\rangle] = 0 \qquad (6)$$

This form of the sum rule was discussed by Nozières and Pines,[5] but can also be found in earlier literature.[6]

It is obvious that the virial theorem and Feynman's theorem have something in common. To find what it is, consider the non-relativistic single-particle Hamiltonian

$$H(\alpha) = \frac{\mathbf{p}^2}{2\alpha} + \sum_{a,b,c} g_{abc} x^a y^b z^c \equiv T + V \qquad (7)$$

where $a + b + c = s$. Dimensional analysis shows that

$$E_n(\alpha) = f(n)\alpha^{-s/(s+2)} h^{2s/(s+2)} G(g) \qquad (8)$$

where G is a homogeneous function of degree $2/(s + 2)$ in the g_{abc}. Planck's constant is denoted by h. It follows that one may write equation (1)

$$s\langle nl|V|nl\rangle = \frac{2s}{2 + s}E_n = -2\alpha\frac{\partial E_n}{\partial \alpha} \qquad (9)$$

This teaches one that the virial theorem also involves a single differentiation of a relation for the energy with respect to a parameter.

For an electron of wave vector $\mathbf{k}$ in a periodic lattice, the reciprocal effective mass tensor can be defined by

$$\left(\frac{1}{m^*}\right)^{n,\mathbf{k}}_{ij} = \hbar^{-2}\frac{\partial^2 E_n(\mathbf{k})}{\partial k_i \partial k_j} \qquad (10)$$

This suggests that the sum rule (5) involves two differentiations with respect to two out of three available parameters. These are the Cartesian components of the wave vector $\mathbf{k}$. Thus, in a general theory one should consider a set of parameters which can be referred to collectively as a parameter vector $\boldsymbol{\alpha}$ (which need not have three components).

One might wish to establish a similar point with respect to (4) and (6). This can be achieved by the artifice of passing from the given Hamiltonian, denoted by H_0 to one into which a parameter vector α has been artificially introduced. Let

$$\mathbf{F} = [F_1(\mathbf{r}_1, \ldots, \mathbf{r}_N), F_2(\mathbf{r}_1, \mathbf{r}_2, \ldots, \mathbf{r}_N), \ldots]$$

where we have an arbitrary number of components, each a suitably chosen function of the $\mathbf{r}$ values. Then we define

$$H_\alpha \equiv e^{-i\alpha \cdot \mathbf{F}} H_0 e^{i\alpha \cdot \mathbf{F}} \qquad u_{nl} \equiv e^{-i\alpha \cdot \mathbf{F}} \psi_{nl} \tag{11}$$

The original Schrödinger equation then leads to an equivalent one with the same eigenvalue, but now incorporating the desired parameters:

$$H_0(\mathbf{p}_1, \ldots, \mathbf{p}_N; \mathbf{r}_1, \ldots, \mathbf{r}_N)\psi_{nl}(\mathbf{r}_1, \ldots, \mathbf{r}_N) = E_n \psi_{nl}$$

$$\Rightarrow H_\alpha(\mathbf{p}_1, \ldots, \mathbf{p}_N; \mathbf{r}_1, \ldots, \mathbf{r}_N)u_{nl}(\alpha; \mathbf{r}_1, \ldots, \mathbf{r}_N) = E_n u_{nl} \tag{12}$$

We now apply this idea to an N-electron system (for instance, an atom) in some velocity-independent field of force. So

$$H_0 \equiv \frac{1}{2m} \sum_1^N \mathbf{p}_j^2 + V(\mathbf{r}_1, \ldots, \mathbf{r}_N) \qquad \mathbf{F} \equiv \sum_{j=1}^N \mathbf{r}_j \tag{13}$$

whence

$$H_\alpha = \frac{1}{2m} \sum_1^N (\mathbf{p}_j + \hbar\alpha)^2 + V \tag{14}$$

It follows that

$$m\hbar^2 \frac{\partial H_\alpha}{\partial \alpha_i \partial \alpha_j} = N\delta_{ij}$$

This suggests that the first terms in both (4) and (5) have a common origin, since $N = 1$ in (5), and that a second derivative is involved. Also, in the atomic case, the quantity analogous to (10) vanishes, since E_n is here independent of the arificially introduced parameter.

Cases **A** to **D** may thus be special cases of a more general theory. This suggests a more systematic discussion.

3. THEORY

Consider an eigenvalue equation

$$A_n(\alpha)\psi_{nl} = 0 \tag{15}$$

with

$$A_n \equiv H(\alpha) - E_n(\alpha)$$

where the given Hamiltonian and eigenvalue depend on some parameter vector.

Differentiate (15) with respect to α_i to find

$$\left\langle n'l'\left|\frac{\partial A_n}{\partial \alpha_i}\right|nl\right\rangle = [E_n(\alpha) - E_{n'}(\alpha)]\left\langle n'l'\left|\frac{\partial}{\partial \alpha_i}\right|nl\right\rangle \tag{16}$$

After some manipulation, a second differentiation leads to

$$-\left\langle nl'\left|\frac{\partial^2 A_n}{\partial \alpha_i \partial \alpha_j}\right|nl\right\rangle = \boxed{2} \sum_{\substack{n''(\neq n) \\ l''}} \frac{\left\langle nl'\left|\frac{\partial A_n}{\partial \alpha_i}\right|n''l''\right\rangle\left\langle n''l''\left|\frac{\partial A_n}{\partial \alpha_j}\right|nl\right\rangle}{E_n(\alpha) - E_{n''}(\alpha)} \tag{17}$$

or, equivalently, to

$$-\left\langle nl'\left|\frac{\partial^2 A_n}{\partial \alpha_i \partial \alpha_j}\right|nl\right\rangle = \boxed{2} \sum_{n'',l''} (E_{n''} - E_n)\left\langle nl'\left|\frac{\partial}{\partial \alpha_i}\right|n''l''\right\rangle\left\langle n''l''\left|\frac{\partial}{\partial \alpha_j}\right|nl\right\rangle \tag{18}$$

The $\boxed{2}$ has the same meaning as in (5), and summations are understood to include integration over the continuous spectrum. Higher differentiations have been carried out, but they lead to more complicated formulae, with which we need not be concerned here. These results are exact, and solely a consequence of (15).

It is perhaps surprising that such a simple theory turns out to be flexible and of wide application. This will now be demonstrated.

We first illustrate (16) by two examples. If one puts $n = n'$ in (16) and remembers that $A_n = H(\alpha) - E_n(\alpha)$, one finds a slight generalization of Feynman's theorem:

$$\left\langle nl'\left|\frac{\partial H}{\partial \alpha_i}\right|nl\right\rangle = \frac{\partial E_n}{\partial \alpha_i}\delta_{l'l} \tag{19}$$

To obtain the virial theorem one must utilize the Hamiltonian (7). Putting this into (19) and utilizing the dimensional analysis (8) yields

$$\langle nl'|T|nl\rangle = \frac{s}{2+s}E_n\delta_{ll'} \tag{20}$$

Since obviously

$$\langle nl'|T + V|nl\rangle = E_n\delta_{ll'} \tag{21}$$

the virial theorem follows in the form (1).

Although these results do not involve sums, it is convenient to call them *sum rules of the first kind*. Sum rules of the second kind are based on (17) and (18) and are the same as what are *conventionally* called sum rules. For example, for the Hamiltonian (7), using (8) and (20), one finds

$$\sum_{\substack{n''(\neq n) \\ l''}} \frac{\langle nl'|T|n''l''\rangle\langle n''l''|T|nl\rangle}{(E_n - E_{n''})} = -\frac{s}{(2+s)^2}E_n\delta_{ll'} \tag{22}$$

This appears to be a new result and extends the virial theorem. The higher order sum rules for this Hamiltonian have been worked out, but become rapidly more complicated.

We now turn to a proof of the effective mass rule from (17). Here a parameter vector $\boldsymbol{\alpha}$ is provided by the wave vector $\mathbf{k}$. Equation (15) is taken in the form

$$H_0 \psi_{nl}(\mathbf{k}, \mathbf{r}) = E_n(\mathbf{k})\psi_{nl}$$

and converted to

$$H_\mathbf{k} u_{nl}(\mathbf{k}, \mathbf{r}) = E_n(\mathbf{k})u_{nl} \tag{23}$$

by using (11) and (12). Note that the E_n and the ψ_{nl} depend on $\mathbf{k}$. Since in the right-hand side of (17) only nondiagonal elements occur, the terms in $\partial E_n/\partial k_i$ do not contribute, so that

$$\frac{\partial A_n}{\partial \alpha_i} \to \frac{\partial H_\mathbf{k}}{\partial k_j} = ie^{-i\mathbf{k}\cdot\mathbf{r}}[H_0, x_j]e^{i\mathbf{k}\cdot\mathbf{r}} = he^{-i\mathbf{k}\cdot\mathbf{r}}v_j e^{i\mathbf{k}\cdot\mathbf{r}} \tag{24}$$

The exponentials disappear on returning to the ψ-representation. The jth component of the electron coordinate has been denoted by x_j.

The second derivative term in (18) contains the reciprocal effective mass tensor by (10). This leaves

$$\frac{\partial^2 H_\mathbf{k}}{\partial k_i \partial k_j} = -e^{-i\mathbf{k}\cdot\mathbf{r}}[[H_0, x_i], x_j]e^{i\mathbf{k}\cdot\mathbf{r}} \tag{25}$$

For any Hamiltonian of the form

$$H_0 = \frac{p^2}{2m} + \mathbf{f}(\mathbf{r})\cdot\mathbf{p} + g(\mathbf{r}) \tag{26}$$

equation (25) yields δ_{ij}/m. Thus, (17) implies (5). If spin-orbit interaction is neglected, the term in $f(\mathbf{r})$ is zero. One then finds

$$\mathbf{v} = \frac{\mathbf{p}}{m}$$

but this relation is not true in general.

4. ARTIFICIAL INTRODUCTION OF PARAMETERS

In the last section, parameters actually occurring in the given Hamiltonian have been used as the parameter $\boldsymbol{\alpha}$ of the general theory. If the mass is used, the virial theorem is obtained as well as the result (22). Many others can also be derived, but they cannot be discussed here. If the wave vector k is used, the effective mass rule is found. Even though parameters (such as a mass) are always present in a Hamiltonian, still it is often desirable to introduce an additional parameter α in order that these may bring in useful new functions. This point will now be illustrated by two examples of many-particle systems. The eigenvalues are in such cases independent of α, so that A_n in (17) and (18) can be replaced by H_α.

To prove the Thomas-Kuhn sum rule it is most convenient to use the F-function and the Hamiltonian given by (13). The basic equation (15) will be taken in the second form (12),

$$A_n \psi_{nl} = 0 \Rightarrow H_\alpha u_{nl}(\alpha) = E_n u_{nl}(\alpha) \tag{27}$$

where H_α is given by (14). Hence,

$$\frac{\partial A_n}{\partial \alpha_i} \to \frac{\hbar}{m} \sum_{w=1}^{N} (p_{w_i} + \hbar\alpha_i)$$

$$\frac{\partial^2 A_n}{\partial \alpha_i \partial \alpha_j} \to \frac{N\hbar^2}{m}\delta_{ij} \tag{28}$$

and (17) yields (with $l = l'$ and after multiplying by $m/\hbar^2$)

$$N\delta_{ij} + \boxed{2}{m} \sum_{\substack{n''(\neq n)\\l''}} \frac{(nl|\sum_w(p_{wi} + \hbar\alpha_i)|n''l'')(n''l''|\sum_w(p_{wj} + \hbar\alpha_j)|nl)}{E_n - E_{n''}} = 0 \tag{29}$$

The round brackets are a reminder that matrix elements have been calculated with respect to the u rather than the ψ of (11). But the switch back from the u- to the ψ-representation is immediate, and one finds a slight generalization of the required sum rule (4).

If one uses (18) in conjunction with the transformation (12), A_n can be replaced by H_α and $\partial/\partial\alpha_j$ by $-iF_{ij}$. If we make this substitution in (18), one finds

$$-\left(nl'\left|\frac{\partial^2 H_\alpha}{\partial \alpha_i \partial \alpha_j}\right|nl\right) = \boxed{2}\sum_{n'',l''}(E_n - E_{n''})(nl'|F_i|n''l'')(n''l''|F_j|nl) \tag{30}$$

The operator occurring on the left is

$$-e^{-i\alpha\cdot\mathbf{F}}[[H_0, F_i], F_j]e^{i\alpha\cdot\mathbf{F}} \tag{31}$$

Inspection now shows that the Nozières-Pines rule (6) can be obtained by using a two-component $\mathbf{F}$:

$$\mathbf{F} = (\rho_\mathbf{k}^+, \rho_\mathbf{k}) \tag{32}$$

and noting that for the Hamiltonian $(4')$

$$[[H_0, \rho_k^+], \rho_k] = -\frac{N\hbar^2 k^2}{m} \tag{33}$$

A return to the ψ-representation is again immediate.

5. CONCLUSION

The considerations advanced here show that given any system, specified by a Hamiltonian, an infinity of sum rules exist. The pattern set by the general forms (16), (17), and (18) can be used to suggest most suitable procedures in any given case.

Among the new results obtainable, and now under study, are those appropriate to more general Hamiltonians than have been considered here. In particular, the restriction to velocity-independent forces can be lifted. The approximate relativistic single-particle Hamiltonians of interest in band theory have also certain simple sum rules. This is not obvious, since one of the approximate forms involves a term in p^4. These situations will be considered elsewhere.

We list two sources[7,8] which are closely related to this work, but could not be explicitly discussed here. We have investigated the connection with perturbation theory elsewhere.[9]

REFERENCES

1. H. A. Bethe and E.E. Salpeter, *Quantum Mechanics of One- and Two-Electron Atoms*, Springer-Verlag, Berlin, 1957, p. 251.
2. R. P. Feynman, *Phys. Rev.* **56**: 340 (1939).
3. Bethe, *op. cit.*, p. 342.
4. E. I. Blount, in *Solid State Physics*, Vol. 13, Academic Press, New York, 1962.
5. P. Nozières and D. Pines, *Phys. Rev.* **109**: 741 (1958).
6. L. D. Landau and E. M. Lifshits, *Quantum Mechanics*, Pergamon Press, London, 1958, p. 460.
7. J. C. Y. Chen, *J. Chem. Phys.* **40**: 615 (1964) and references cited therein.
8. A. Dalgarno, *Rev. Mod. Phys.* **35**: 522 (1963) and references cited therein.
9. P. T. Landsberg and D. J. Morgan *J. Math. Phys.*, December 1966 (to be published).

Inelastic Neutron Scattering and Dynamics in Solids and Liquids

ALF SJÖLANDER

CHALMERS TEKNISKA HÖGSKOLA, INSTITUT FÖR MEKANIK
Göteborg, Sweden

1. INTRODUCTION

The concept of lattice waves, called *phonons,* propagating through single crystals goes back to Debye and Born-von Kármán (1912). In fact, it can be traced back even further. The phonons appear as the elementary excitations of the lattice, and each one is characterized by its wave vector, frequency, and polarization direction. In the simplest case, where we have only one atom per unit lattice cell, there are for each wave vector three different phonons, one essentially longitudinal and two essentially transverse. These are called *acoustic phonons.* For more than one atom per unit cell, there are a greater number of phonons for a given wave vector. In the alkali halides, which contain two atoms per unit cell, we have three acoustic branches and three other branches, called *optical phonons.* They represent motions where neighboring atoms move out of phase even for long-wavelength vibrations and thus give rise to an oscillating electric dipole moment within each unit cell. For this reason, the very-long-wavelength optical phonons are responsible for strong infrared absorption in alkali halides. The long-wavelength acoustic phonons affect the elastic properties of the crystal, and the short-wavelength acoustic phonons determine the thermodynamic properties of the lattice—that is, the free energy, specific heat, entropy, etc.

The theories of Debye and Born–von Kármán were quite successful in explaining various experimental results on thermal motions in a lattice. However, it was not until much later that experimentalists were able to demonstrate the existence of these phonons more directly. This was first done by measuring the weak diffuse scattering of X-rays by a single crystal. In this connection, the measurements by Jakobsen on copper and Walker on aluminum should be mentioned. By carrying out very detailed and quite elaborate analyses of their experimental data, they obtained the dispersion curves for the phonons (that is, the phonon frequency as a function of the wave vector) in the symmetry directions of the crystal.

At present, experimental determinations of phonon dispersion curves and polarization directions are exclusively done by inelastic neutron scattering. It will be apparent in the following discussion why this technique is superior to that of X-ray scattering. In some cases, the weak part of the infrared absorption, as well as second-order Raman scattering, give quite accurate information on the frequencies of certain phonons.

However, the usefulness of the phonon concept becomes more and more doubtful as the melting point of the crystal is approached. The phonons, if they exist, should be strongly damped because of anharmonic effects, and we should expect the motion of the atoms to be drastically changed above the melting point. The long-wavelength motion would then be governed by the macroscopic hydrodynamic equations, which certainly are quite different from the equations appearing in the theory of elasticity. So, for instance, there are no long-wavelength transverse waves in a liquid. Scattering of light, where the wavelength considered is of the order 10^4Å, does reveal in some cases that the Navier-Stokes equations need to be modified for shorter wavelengths.

In neutron scattering, the wavelength region covered is of the order 1 to 20 Å. The thermal fluctuations in a liquid are normally in the same wavelength range, and, therefore, inelastic neutron scattering is particularly suitable for determining the thermal motions in liquids.

2. INELASTIC NEUTRON SCATTERING TECHNIQUE

In principle, what one does in neutron scattering experiments is to let a beam of well-collimated monochromatic neutrons pass through a sample of a single crystal, or some other material to be studied, and measure the energy distribution of the neutrons scattered in a specified direction. The incident energy is normally of the order 10^{-3} to 10^{-4} eV, wavelength 3 to 6 Å. There are two different kinds of scattering: If all the scattering centers are identical and no spin flip occurs, interference between scattered neutron waves originating from different nuclei occurs, and therefore the scattering amplitudes add up, giving significant interference effects. This is called *coherent scattering*. If there is spin flip in the scattering between the neutrons and the individual nuclei, and also as a result of isotopic disorder in the sample, an extra scattering, which does not give any interference effects, occurs, called *incoherent scattering*. In this case, the scattering cross sections from the individual nuclei add up.

In the scattering process between the incident neutrons and a single crystal, one, two, or more phonons can be created, leading to a change in the energy and momentum of the neutrons. Particularly at elevated temperatures, phonons present in the sample can be annihilated, leading to an increase of the neutron energy. Scattering where no phonons are created or annihilated is strictly elastic and has long been known as *Bragg scattering*. Processes in which only one phonon is created or annihilated give rise to a very distinct resonance peak in the energy distribution of the scattered neutrons. Higher order processes involving more than one phonon give only a rather smooth background, and can in most cases easily be subtracted along with other kinds of background scattering. The position of the resonance peaks is related to the wave vector and frequency $\omega(\mathbf{q})$ of the phonon involved through the conservation rules for energy and momentum:

$$\mathbf{k}_0 - \mathbf{k} = \mathbf{q} + \boldsymbol{\tau}$$
$$E_0 - E = \pm\hbar\omega(\mathbf{q}) \tag{1}$$

The wave vectors of the incident and scattered neutron are $\mathbf{k}_0$ and $\mathbf{k}$, respectively; E_0 and E are the corresponding energies, where

$$E = \frac{\hbar^2 k^2}{2m} \tag{2}$$

$\boldsymbol{\tau}$ denotes a reciprocal lattice vector and arises from the periodicity of the lattice. We may state that the lattice as a whole can absorb momenta equal to a reciprocal lattice vector times $\hbar$. In this way, the wave vectors of the phonons are always restricted to the first Brillouin zone. The plus and minus signs in the second line of (1) correspond to phonon emission and phonon absorption processes, respectively.

From the experimental arrangement, k_0, k, E_0, and E are obtained directly, and from (1), $\omega(q)$ is thus obtained for a specific value of q. In fact, several resonance peaks, corresponding to the various polarization directions of phonons with the same wave vector, are obtained.

In some favorable cases, it has also been possible to determine the lifetimes of the phonons by measuring the width of the resonance peaks, and this has thrown light on the validity of the phonon concept at higher temperatures.

In a polycrystalline material, the various crystals have different directions, corresponding to rotating τ in (1). Because of this, no distinct resonance peaks appear in the scattered neutron beam. However, certain sharp rises in the energy distribution reflect the presence of single phonons, and in favorable cases some conclusions about phonon dispersion curves can also be drawn from measurements on polycrystalline materials.

In the case of liquids, there is no dynamical theory to describe the atomc motions, and at present the only recourse is to use very simplified models. It is particularly for liquids that inelastic neutron scattering seems a promising method for getting a more detailed picture of the atomic motions than that which can be obtained by other means. Unfortunately, the effects are expected to be even less pronounced than those found in polycrystalline materials. For this reason, analysis of experimental data is very difficult in the absence of a proper theory.

In X-ray diffraction experiments, it has not been possible to measure the energy change of the scattered photon directly, because of insufficient energy resolution, which prevents any measurement of resonance effects. It is necessary to measure an integrated intensity, corrected for various effects, and finally to compare this with a detailed theory. The corrections are difficult to carry out with high accuracy, and so phonon dispersion curves of sufficient accuracy become very difficult to obtain. In a polycrystalline material, this becomes impossible, and information on phonon lifetimes cannot be obtained.

The recent discovery of the Mössbauer effect gives some hope that one should be able to carry out scattering experiments with X-rays where both the momentum and energy changes of the photon can be measured, as they are for the neutrons in neutron scattering experiments. So far, the intensity has been very low, but the energy resolution is far better than that obtained with neutron scattering.

3. THEORY OF INELASTIC SCATTERING OF NEUTRONS

The most general and at the same time the most suitable formulation of the theory for slow neutron scattering was given by Van Hove (1954). He showed that the incoherent and coherent differential scattering cross sections could be expressed as space–time Fourier transforms of certain correlation functions, assuming the scattering system to be in thermal equilibrium:

$$\frac{d^2\sigma(\text{inc})}{d\Omega d\omega} = Nb_{\text{inc}}^2 \frac{k}{k_0} \frac{1}{2\pi} \int_{-\infty}^{\infty} e^{-i\omega t} F_s(\kappa t) dt \qquad (3)$$

$$\frac{d^2\sigma(\text{coh})}{d\Omega d\omega} = Nb_{\text{coh}}^2 \frac{k}{k_0} \frac{1}{2\pi} \int_{-\infty}^{\infty} e^{-i\omega t} F(\mathbf{k}t) dt \qquad (4)$$

where

$$\kappa = \mathbf{k}_0 - \mathbf{k} \qquad \hbar\omega = E_0 - E \qquad (5)$$

here mean the respective change in the wave vector and energy of the scattered neutron, N is the number of nuclei in the scattering sample, assuming a monatomic system, and k/k_0 gives the ratio between the velocity of the scattered neutron and that of the incident neutron; "b_{inc}" and "b_{coh}" are certain constants which depend only on the kind of nucleus scattering the neutrons.

The dynamics of the scattering system is confined to the two basic functions $F_s(\kappa, t)$ and $F(\kappa, t)$, defined as the following correlation functions, averaged over an equilibrium ensemble:

$$F_s(\mathbf{k}, t) = \langle \exp[-i\mathbf{k}\cdot\mathbf{r}_0(0)] \exp[i\mathbf{k}\cdot\mathbf{r}_0(t)] \rangle \qquad (6)$$

$$F(\kappa, t) = \sum_l \langle \exp[-i\kappa\cdot\mathbf{r}_0(0)] \exp[i\kappa\cdot\mathbf{r}_l(t)] \rangle \qquad (7)$$

Here, $\mathbf{r}_0(t)$ means the position vector operator for an arbitrary nucleus, given above the index 0. In the second definition, we take into account the position of two nuclei at different times, and the summation extends over all nuclei. We see that the time evolution of the scattering system will determine how the neutrons are scattered.

We shall not go further into the theory. I shall only mention that a great many authors have tried to draw various conclusions from this general formulation. It is indeed possible to make some very general and important statements about the scattering; and in other cases, certain models have been assumed—for instance, Born-von Kármán equations for a harmonic solid—with which more specific conclusions have been reached.

As one illustration, I should mention that it follows directly from the formulation that, neglecting diffusion effects in a solid, a finite fraction of the scattering is strictly elastic and the intensity of the elastic part is governed by the Debye-Waller factor

$$(\langle e^{-i\kappa \cdot r_0}\rangle)^2 \tag{8}$$

The conservation rules stated in (1) follow directly from the formulation above, assuming a harmonic single crystal.

4. RESULTS ON PHONONS IN SOLIDS

At present, a large amount of experimental data on phonon dispersion curves for various crystals at various temperatures has been collected. It is quite remarkable that even up to temperatures fairly close to the melting point, distinct single phonon peaks have appeared. The broadening of the resonance lines becomes significant. In aluminum, the phonon lifetime near the melting point has been found to be of the order of two periods. This is, however, enough to give a clear resonance peak.

It has in many cases been difficult to explain in detail the experimental dispersion curves on the basis of the Born-von Kármán theory, in which a short-range interatomic interaction is essentially assumed, since it turns out that in metals the conduction electrons give rise to a fairly long-range effective interaction between the atoms. Polarization of the more tightly bound electrons also gives a long-range interatomic interaction. For these reasons, more sophisticated theories based on concepts from many-body theory have been worked out, and the results obtained seem quite promising. However, further theoretical work must be done.

Particularly in recent years, extensive study of effects from anharmonicity has been made as, for instance, phonon life-times. A difficulty here is the lack of information on the anharmonic force constants, and no detailed comparison of theory with experimental results has been made.

Regarding the phonon polarization directions, only very few experimental data have been reported so far, and it is mainly in the symmetry directions of the crystals that the phonon dispersion curves have been obtained.

5. RESULTS ON LIQUID DYNAMICS

In several respects, the study of liquids is more challenging than the study of solids. Inelastic neutron scattering seems to provide a unique opportunity to learn both about how single atoms move in liquids and how neighboring atoms are correlated in their motions. As stated previously for the case of polycrystalline materials, analysis of the experimental data is far from straightforward, since no resonance peaks appear. There is no dynamical theory for a liquid capable of serving the same purpose as, for instance, the Born-von Kármán theory for solids. Many hypotheses regarding the thermal motions in a liquid have been put forth over the past years, many of which go back to Frenkel. From data on the thermodynamic and transport properties of liquids, Frenkel argued that the motion of the atoms should be similar to that in solids, assuming, of course, the proper consideration of diffusion effects. This theory would seem particularly valid at temperatures not too far from the melting point; however, Frenkel had no direct evidence for these assumptions.

Neutron scattering experiments give direct information on the motion of a single atom. Before any detailed experiment on liquids had been performed, it was suggested that the atom might be treated as a Brownian particle, following Langevin's equation

$$m\dot{\mathbf{v}} + m\zeta\mathbf{v} = \mathbf{F}(t) \tag{9}$$

where $\mathbf{v}(t)$ stands for the velocity vector of the atom, and m is its mass; ζ is a phenomenological friction constant, related to the macroscopic self-diffusion constant as follows:

$$D = \frac{k_B T}{m\zeta} \tag{10}$$

with k_B the Boltzmann constant, and T the temperature. $\mathbf{F}(t)$ is a random fluctuating force, which is assumed to have a white spectrum. From this model, it is very easy to determine the scattering cross section; the result is essentially

$$\frac{d^2\sigma(\text{inc})}{d\Omega\,d\omega} = Nb_{\text{inc}}^2\frac{k}{k_0}\left[\frac{1}{\pi}\frac{D\kappa^2}{\omega^2 + (D\kappa^2)^2}\right] \tag{11}$$

There are three particular conclusions we can draw from this formula:

1. For a fixed value of κ, the last factor has a Lorentzian form.
2. The width of the Lorentzian curve is given by

$$\Delta\omega = 2D\kappa^2 \tag{12}$$

3. The integrated value over the frequencies is unity.

Later on, calculations based on a very different model were presented, in which it was essentially assumed that the atoms are for some time trapped by the surrounding atoms and make oscillatory motions similar to those in a solid. Now and then, the atoms go over into diffusive motions and after a short time get trapped again. This resembles the diffusion mechanism in a solid. One essential time enters into this model, the average time the atom is trapped before making a diffusive jump. We shall call that time τ_0. The scattering cross section consists of two different parts. One part refers to the diffusion effects and has the form

$$\left[\frac{d^2\sigma(\text{inc})}{d\Omega\,d\omega}\right]^{(1)} = Nb_{\text{inc}}^2\frac{k}{k_0}\left[\frac{1}{\pi}e^{-a\kappa^2}\frac{f(\kappa)}{\omega^2 + [f(\kappa)]^2}\right] \tag{13}$$

where

$$f(\kappa) = \frac{1}{\tau_0}\left(1 - \frac{e^{-a\kappa^2}}{1 + D\kappa^2\tau_0}\right) \tag{14}$$

The second part refers to the oscillatory motion and gives rise to larger energy transfers, which are similar to those in solids. For this reason, the first part gives a quasi-elastic scattering peak with a width given by $f(\kappa)$, and the second part gives the wings of this peak. The main conclusions which may be drawn from this model are as follows:

1. For a fixed value of κ, the quasi-elastic peak is Lorentzian.
2. The width of this quasi-elastic peak is given by a function $f(\kappa)$, which for $D\kappa^2\tau_0 \ll 1$ goes over into

$$\Delta\omega = 2D\kappa^2 \tag{15}$$

and for $D\kappa^2\tau_0 \gg 1$ goes over into

$$\Delta\omega = \frac{2}{\tau_0} \tag{16}$$

3. The integrated intensiy of the quasi-elastic peak is given by a factor

$$e^{-a\kappa^2} \tag{17}$$

where a is essentially the mean square amplitude for the oscillations when the atom is trapped. This intensity factor plays exactly the same role as the Debye-Waller factor does in solids.

4. The wings of the scattering peak should extend to essentially the same energy transfers as those found in corresponding solids.

The first experiments were done at the same time that this second model was proposed, and were carried out for ordinary water. It was clearly shown that Langevin's equation did not adequately describe the experimental results, which gave a width of the quasi-elastic peak an order of magnitude smaller than that given in equation (12). However, it could be explained on the basis of equation (14) by choosing $\tau_0 = 3 \times 10^{-12}$ sec. This time is also consistent with some other kinds of experiments. Later experiments have shown that the width varies with κ essentially as given by equation (14). The integrated intensity is also consistent with (17), assuming an effective Debye temperature for water of around 130°K. All these results refer to room temperature. Nowadays, a quite extensive collection of experimental data for various liquids has been obtained, and they seem in general to favor the second model. However, they also show that refinements are needed, since the actual motion seems to be somewhere between Langevin-type diffusion and jump diffusion. Jump diffusion is more pronounced for very viscous liquids such as glycerol.

Coherent neutron scattering should reveal correlation in the motion of neighboring atoms; however, very little theoretical work has been done, and experimental work has been limited to obtaining only preliminary data. However, these studies are extremely interesting and revealing. Scattering from a polycrystalline sample of aluminum below the melting point has been compared with scattering from liquid aluminum close to the melting point. It was found to be barely possible to distinguish the two scattering curves, in spite of the fact that quite a lot of the details of the atomic motion enters into the scattering cross section. Similar results were obtained for lead. In argon, on the other hand, there is not this striking similarity. This seems to show quite conclusively that for a distance of up to 10 to 20 Å the atomic motion in many liquids is very much the same as in solids. There is a strong correlation extending at least over this distance.

There is at present great interest in exploring coherent scattering in liquids, and from the theoretical standpoint there is, of course, great interest in devising suitable models. It is unfortunate that because of insufficient resolution, neutron scattering cannot at present be used to obtain information on atomic motions for times longer than 10^{-12} to 10^{-11} sec and distances greater than about 20Å. The essential relaxation times seem to be of the order 10^{-11} sec, and the correlation distance of the order 20 Å. Scattering of light, on the other hand, gives information only for motions over a distance of more than 1000 Å and for times longer than 10^{-10} sec. It is of particular interest to fill that gap, and it may be that Mössbauer scattering experiments will do that work. Efforts are also made to improve the neutron experiments, going to lower incident energies and smaller scattering angles in order to reach longer times and greater distances.

In discussing correlations, I have been very vague, since at present we do not really know exactly what "correlation" means. Precise definitions are greatly needed, and probably the best way to get them is to introduce some correlation functions as is done in many-body theory. In addition, equations of motion for these correlation functions must be derived or postulated and they must be at least approximately solvable. At present, we are certainly very far from this goal.

Axioms and Models

M. H. STONE

UNIVERSITY OF CHICAGO
Chicago, Illinois

The connection between physics and mathematics poses a problem which has both practical and philosophical aspects. It has not seemed necessary, except perhaps to philosophers, to discuss this problem in any detail. However, the increasingly mathematical character of physics (and of other branches of science, we should add) draws our attention more and more urgently to this problem.

The fashionable way of treating the problem is to say that the physicist (or the biologist, economist, or pyschologist) constructs a "mathematical model" of the phenomenon under investigation. The problem is then discussed with a few rather superficial observations about abstraction, idealization, and approximation. The inquiry is rarely pressed far enough to disclose its logical and metaphysical roots or to clarify and enrich the techniques of model construction.

The state of elementary particle physics, as discussed in such a masterly way by Professor Weisskopf in his Introductory Address, suggests that a close examination of the problem would not be without value. The emergence (for different particle spectra) of a common pattern, rooted in underlying symmetries, has been accompanied by the development of more and more explicit techniques for constructing models based on group theory. In the background lie earlier techniques developed for treating the dynamics of simple and composite physical systems. These techniques are based on the theory of linear spaces (especially Hilbert space) and their spectral resolution and synthesis. In fact, the theory of group representations emerges from coupling the theory of abstract groups with the theory of linear spaces. The fairly successful attack on the problems of representation theory (which began long before the quantum theory was thought of, but which was greatly stimulated by the ideas of modern physics) provides some techniques for finding suitable models. Mathematicians have perhaps had greater confidence than physicists in these techniques, because since the early thirties some of them have had in mind a program of looking for the groups which would possibly have uses in physics and trying them out in a systematic way. However, the researches of mathematicians directed toward carrying out this program have been turned away from questions of utility for physics by the difficult mathematical obstacles which had to be overcome (particularly in relation to infinite-dimensional representations). The most striking current applications $[SU(3)$ and $SU(6)]$ were proposed by theoretical physicists rather than by the mathematicians interested in the general program.

Any study of model construction has to begin with a logical analysis of the procedures involved. This part of the investigation can be considered as having been very well mapped out by logicians and mathematicians. At bottom, model construction is an exercise in axiomatics. The specifications for a model are the axioms of a logico-mathematical theory, and the highly developed metatheory of logico-mathematical theories and their axiomatic presentation can be brought to bear. The essential point is that a given set of propositions (the axioms) generates a deductive system, whose elements are the propositions (theorems) deducible from the axioms. The specifications for a model are the axioms for such a deductive system. It is evident that the same body of theorems may be generated by different bodies of axioms. This possibility challenges the logician, the mathematician, and the scientist to investigate the body of axioms on which a theory is to be based and to select the one most convenient for this purpose. While there are techniques for doing this, intuition (whether logical, mathematical, or physical) is most important in practice. A logico-mathematical theory has in the nature of things an abstract symbolic form, which often needs to be given a concrete interpretation or representation. This is the "model" that physicists customarily speak of. In many cases, the passage from theory to model is made by the introduction of parameters or coordinates specifying abstract configurations of the abstract system. Thus, a model is given by specifying its basic abstract properties and suitable coordinates. For a given deductive theory there may be many different models.

The first example of a deductive system was provided long ago by the Greek geometers. Euclid brought together their results in his famous *Elements*, writing down a body of axioms and presenting the deductions of a large number of theorems, some of which had long been known empirically and many of which, we now realize, represent steps in an algebraic inquiry into properties of space. It was a long time before Descartes produced, in his analytic geometry, a model for this deductive system. In a very roundabout way Descartes saw that the axioms of Euclid involved the presence of axioms for a number system for which a model could be produced. Today, the relation between geometry and number theory is much more clearly understood, but the result is the same: a deductive geometry contains the material for its own coordinatization in terms of a number system. The number system itself has to be specified by axioms, and a concrete realization of it is a model in terms of which a model for the geometry itself can be produced. Underlying this connection between geometry and numbers, there is a notion of symmetry, not brought out explicitly in Euclid's axioms, but now treated properly in all modern discussions with any pretense at accuracy and completeness. In the nineteenth century the mathematician Felix Klein had already formulated the general principle that each geometry arises out of symmetries defined by a specific group. This principle has been a guide to geometers ever since. The principle leads eventually to the identification of all geometrical configurations (for example, points, lines, and planes in Euclidean geometry) as configurations in the underlying characteristic group.

The use of the term "model" is frequently intended to emphasize the distinction between the phenomenon under discussion and the model for it, not merely the difference between a deductive system and its interpretation as a model. Furthermore, it is just as frequently intended to emphasize that the model may deviate in certain respects from what it is intended to represent. It is perhaps helpful to resolve the passage from phenomenon to model, as we have implicitly done here, into two steps—the passage from the phenomenon to the specifications or axioms with their attached deductive system, followed by the passage from this deductive system to a model for it.

It can then be seen quite clearly that the axioms or specifications are in the nature of carefully phrased stipulations as to what we know (or guess) and will take into account about a particular phenomenon under investigation. It is evident that certain parts of our information may be deliberately suppressed in order to arrive at simpler and more manageable theoretical models. In scientific practice it is quite usual to modify or extend the stipulations in the course of an investigation. This means changing the axioms and restricting or changing the possible models, and is often done under the spur of necessity, when the mathematical discussion runs up against technical difficulties. A physicist is on much safer ground than a mathematician when he "changes his axioms in midstream," because he has knowledge of a physical situation to guide him. A mathematician, however, can be guided only by mathematical intuition, and is perhaps more likely to fix on a modification leading to inconsistencies. In any case, the effect of modifying the axioms with which one works can be understood in general logical terms and is no more mysterious than logic itself.

A rather notable feature of axioms and models in physics is the extent to which they arise out of the stipulation of what appear to be metaphysical or psychological principles prior to any attempt at grasping the phenomena with which physics is concerned. Already in geometry the basic concepts of point, line, and surface have a most tenuous relation to any physical phenomenon. To deal with them, it is necessary to bring to bear concepts of a logical nature, those from the logic of propositions (proof theory), and those from set theory (set, mapping, etc.). In finding a model for the deductive theory of Euclidean geometry, it is necessary to pass to an axiomatic treatment of numbers and eventually to construct a model of the theory of real numbers in terms of *infinite* sequences of natural numbers (as they emerge from set theory). In the same way, the basic stipulations about quantum physics involve the concepts of "state" and "observable" and a statistical theory of the measurement of "observables." It is hard to escape the feeling that the ideas of Plato and Kant about a priori knowledge are being demonstrated in action by the modern physicist! The alternative is to take a strictly operational point of view, excluding appeal to the notion of infinity (and hence to the real number system) and dealing only with finite sets of natural numbers produced by measuring instruments. The discernment of patterns in such systems would appear, however, to lead to imbeddings in larger systems in which the real numbers reappear. This, of course, is exactly what happens in Euclidean geometry, the appearance of the real numbers not being guaranteed until the "principle of continuity" is invoked as an axiom and the Euclidean plane (or space) "completed." Thus, it may be said that some of the axioms and models of physics are in fact axioms and models for the mind. If, as I have tried to maintain elsewhere, the mind is to be regarded as the principal instrument by which we perceive the phenomena of the physical world, this means only that there is no theory or model of the physical world which ignores this instrument of observation. I consider it more than likely that as we find better theories and models for the mind they will enter explicitly into the theories and models physicists try to construct for the physical world. If there is any clue to the very widespread emergence of symmetries and groups in the treatment of natural phenomena, it is very likely to be found in the way in which the brain functions. A mathematician would feel that in any mathematical system we should look for the underlying symmetries and make clear the role played by the corresponding group. Ultimately, he might expect to reformulate the axiomatic theory of the system in terms of this group. But is not physics today moving precisely in this direction so far as the axioms and models of the physical world are concerned? And may not the group itself be the underlying psychophenomenological reality?

It is impossible now to avoid a very fundamental question: Is the physicist occupied merely with constructing a mathematical model which has necessarily to be considered as an approximation to an ideal model of the material world? It seems to me that, on the contrary, he seeks an exact logical analysis of all he knows about the physical world, in terms of a complete deductive system of propositions embodying his knowledge and its implications. To be sure, he may not at any given moment know all that there is to be known about the physical universe, and he may very well use different models, including approximate models, in detailed studies. This does not alter the fact that his main objective is the one described above. The addition of new facts to the physicist's knowledge forces a revision of his theory and a change of models. Thus, the basic question raised by these remarks is: Can the physicist eventually arrive at a complete knowledge of the physical world? An intriguing possibility is that the univers eis built on an infinite regression of particles, each type of particle being composed, in some sense, of others of a more "elementary" character. Since this is mathematically conceivable, it cannot be ruled out. In practice, such a situation would open up the way to a never-ending successive exploration of the different levels of particles and theoretical attempts to extrapolate to the levels still beyond the range of experiment. On the other hand, it is just as conceivable that the search for elementary particles may come to an end with the discovery of particles of which all matter is built. However, it could happen that any one of these particles might appear as a composite of others at the same level. The theoretical discussion opened up by these speculations may well involve an appeal to the new mathematical theory of categories. This theory, designed originally to disclose and treat the general structure of group theory and of other mathematical theories in a unified way, could conceivably provide valuable techniques in physics.

Characters of Semi-Simple Lie Groups

HARISH-CHANDRA

INSTITUTE FOR ADVANCED STUDY
Princeton, New Jersey

Let C be the circle group, that is, the multiplicative group of all complex numbers c with $|c| = 1$. Then for every integer n, we have the character χ_n of C given by $\chi_n(c) = c^n$, and these are the only (irreducible) characters of C. Moreover, the main result in the theory of Fourier series asserts that

$$f(1) = \sum_{-\infty < n < \infty} \int_0^{2\pi} f(e^{i\theta}) e^{in\theta} \frac{d\theta}{2\pi}$$

for any smooth function f on C. Now $dc = d\theta/2\pi$ may be regarded as the normalized Haar measure on C, so that the total measure of C is 1. Then this equation becomes

$$\delta(f) = \sum_n \int f \chi_n \, dc$$

where δ is the Dirac measure on C concentrated at 1. We may restate this equation in the form

$$\delta = \sum_n \chi_n$$

in the sense of the Schwartz theory of distributions. This relation is called the *Plancherel formula* for C.

Now let G be a compact topological group and $\mathscr{E}$ the set of all equivalence classes of finite-dimensional irreducible unitary representations of G. If π is such a representation, we define

$$\Theta_\pi(x) = \mathrm{Tr}\, \pi(x) \qquad (x \in G)$$

The function Θ_π is called the *character* of π. It depends only on the class ω of π in $\mathscr{E}$. Hence, we may denote it by Θ_ω. Then the mapping $\omega \rightarrow \Theta_\omega$ is one-to-one, and $d(\omega) = \Theta_\omega(1)$ is the degree of any representation in ω. In this case, the Plancherel formula states that

$$\delta = \sum_{\omega \in \mathscr{E}} d(\omega)\Theta_\omega$$

When G is Abelian, $d(\omega) = 1$ for all ω, and this reduces to

$$\delta = \sum_{\omega \in \mathscr{E}} \Theta_\omega$$

This is just what we had above for $G = C$.

However, when G is not compact, it need not have any finite-dimensional *unitary* representation at all, except the trivial one whose kernel is the whole of G. Therefore, we are forced to consider infinite-dimensional representations.

Let G be a connected semi-simple Lie group. (For our purposes, *semi-simple* means that every normal Abelian subgroup of G is finite.) A representation π of G on a Hilbert space $\mathscr{H}$ is a mapping which assigns to every $x \in G$ a unitary operator $\pi(x)$ on $\mathscr{H}$ such that (a) $\pi(xy) = \pi(x)\pi(y)$, $\pi(1) = I$, (b) the mapping $(x, \psi) \rightarrow \pi(x)\psi$ of $G \times \mathscr{H}$ into $\mathscr{H}$ is continuous. π is called *irreducible* if no closed subspace of $\mathscr{H}$, other than $\{0\}$ and $\mathscr{H}$, is stable under $\pi(x)$ for all $x \in G$.

Now fix an irreducible (unitary) representation π of G. We wish to define

$$\text{Tr } \pi(x) \qquad (x \in G)$$

Let us choose an orthonormal base ψ_i $(i = 1, 2, \ldots)$ in $\mathscr{H}$. We would like to define

$$\text{Tr } \pi(x) = \sum_{i \geq 1} [\psi_i, \pi(x)\psi_i]$$

However, the difficulty is that the series does not converge in general. For example, if we take $x = 1$, we get

$$\sum_{i \geq 1} |\psi_i|^2 = \sum_{i \geq 1} 1 = \infty$$

provided dim $\mathscr{H} = \infty$. Now put $\phi_i(x) = [\psi_i, \pi(x)\psi_i]$. Then ϕ_i is a continuous function on G and

$$\sum_i \phi_i$$

regarded as a series of functions, does converge in the space of distributions.

Let $C_c^\infty(G)$ denote the space of all indefinitely differentiable functions f which vanish outside some compact subset of G. Also, let dx denote the Haar measure on G. (Apart from a constant factor, dx is uniquely determined by the condition that

$$\int f(yx)\,dx = \int f(x)\,dx \qquad (y \in G)$$

for any continuous function f on G with compact support.) Then one can show that

$$\sum_{i \geqslant 1} \left| \int f(x)[\psi_i, \pi(x)\psi_i]\,dx \right| < \infty$$

for any $f \in C_c^\infty(G)$. Put

$$\Theta_\pi(f) = \sum_{i \geqslant 1} \int f(x)[\psi_i, \pi(x)\psi_i]\,dx$$

Then Θ_π is a linear function on the vector space $C_c^\infty(G)$. Such functions, provided they satisfy some mild continuity conditions, have been named "distributions" by Schwartz. One can verify that Θ_π is, in fact, a distribution, and we shall call it the character of π.

We now enumerate some properties of Θ_π. Let $\mathscr{E}$ be the set of all equivalence classes of irreducible unitary representations of G.

1. Θ_π *is independent of the choice of the orthonormal base. More-over it depends only on the class* ω *of* π *in* $\mathscr{E}$. *Hence, we may denote it by* Θ_ω. *The mapping* $\omega \to \Theta_\omega$ *is one-to-one.*

Write $y^x = xyx^{-1}$ $(x, y \in G)$. For any function f on G define f^x to be the function given by

$$f^x(y) = f(x^{-1}yx) = f(y^{x^{-1}})$$

f is called invariant if $f^x = f$ for all $x \in G$. Similarly let T be a distribution on G. Then T^x is the distribution given by

$$T^x(f) = T(f^{x^{-1}}) \qquad [f \in C_c^\infty(G)]$$

Again we say that T is invariant if $T^x = T$ for all $x \in G$. Then the second property of Θ_ω may be stated as follows.

2. *For any* $\omega \in \mathscr{E}$, *the distribution* Θ_ω *is invariant.* This is an obvious generalization of the well-known fact that $\operatorname{Tr} \pi(y^x) = \operatorname{Tr} \pi(y)$ for any finite-dimensional representation π.

Now we come to the third and the most important property of Θ_ω. Let D be a differential operator on G. Then its adjoint D^* is the unique differential operator such that

$$\int f D^* g\,dx = \int Dfg\,dx$$

for all $f, g \in C_c^\infty(G)$. If T is a distribution, then the distribution DT is defined by

$$(DT)(f) = T(D^*f) \qquad [f \in C_c^\infty(G)]$$

Fix $x \in G$ and, for any $f \in C_c^\infty(G)$, define the function f_x by $f_x(y) = f(yx)$. We say that f_x is obtained from f by right translation by x. Moreover D is said to commute with right translations if

$$(Df)_x = Df_x$$

For all $f \in C_c^\infty(G)$ and $x \in G$. Similarly one can speak about commuting with left translations.

Let $\mathfrak{Z}$ be the algebra of all differential operators on G which commute with both left and right translations. It is known that $\mathfrak{Z}$ is Abelian.

3. Θ_ω *is an eigendistribution of* $\mathfrak{Z}$. This means that for every $z \in \mathfrak{Z}$, there exists a complex number $\chi(z)$ such that $z\Theta_\omega = \chi(z)\Theta_\omega$.

Let $\mathscr{G}$ be the Lie algebra of G and σ the adjoint representation of G on $\mathscr{G}$. (In case G is a matrix group, $\mathscr{G}$ also consists of matrices and $\sigma(x)X = xXx^{-1}$ for $x \in G$ and $X \in \mathscr{G}$.) Let t be an indeterminate. Then

$$\det[t + 1 - \sigma(x)]$$

is a polynomial in t whose coefficients are analytic functions on G. Moreover the highest coefficient is 1. Let l be the least integer $\geqslant 0$ such that the coefficient $D(x)$ of t^l is not identically zero. A point x in G is called singular or regular according as $D(x)$ is or is not equal to zero. Let G' denote the set of all regular elements. Then G' is open and dense in G and the singular set has Haar measure zero.

THEOREM: *Let* Θ *be an invariant eigendistribution of* $\mathfrak{Z}$ *on G.*

Then there exists a locally summable function F on G which is analytic on G' such that

$$\Theta(f) = \int F \, dx$$

for all $f \in C_c^\infty(G)$. Moreover, F is uniquely determined on G' by these properties.

Now let Θ_π be the character of an irreducible unitary representation of G. Then the above theorem is applicable to Θ_π. Let F_π denote the corresponding function on G. Then

$$\sum_i \int f(x)(\psi_i, \pi(x)\psi_i) \, dx = \Theta_\pi(f) = \int f(x)F_\pi(x) \, dx$$

for $f \in C_c^\infty(G)$. Hence, it seems reasonable to say that

$$\text{Tr } \pi(x) = F_\pi(x) \qquad (x \in G)$$

Of course, in general, the function F_π will have singularities on the singular set. Nevertheless, it will be convenient to define

$$\Theta_\pi(x) = F_\pi(x) \qquad (x \in G')$$

Let us now take an example. Consider the group G of all 2×2 real matrices

$$\begin{pmatrix} a & b \\ c & d \end{pmatrix}$$

with determinant 1 so that $ad - bc = 1$. There are two series of irreducible unitary representations in this case, called the *continuous* and the *discrete series*, respectively.

1. The continuous series is parameterized by a real number λ and for each λ there are two characters T_λ^+, T_λ^-.
2. The discrete series is parameterized by an integer $m \neq 0$ and we denote the corresponding character by Θ_m.

All these characters are distinct and irreducible, except that

$$T_\lambda^+ = T_{-\lambda}^\pm$$

and T_0^- decomposes into two irreducible characters. (We have ignored the exceptional series here, since it does not occur in the Plancherel formula.) Moreover,

$$8\pi f(1) = \sum_{m \neq 0} |m| \Theta_m(f) + \int_0^\infty \frac{\lambda}{2} \tanh \frac{\pi\lambda}{2} T_\lambda^+(f) \, d\lambda$$

$$+ \int_0^\infty \frac{\lambda}{2} \coth \frac{\pi\lambda}{2} T_\lambda^-(f) \, d\lambda$$

for all $f \in C_c^\infty(G)$. This is the Plancherel formula for G.
Put

$$h_t = \begin{pmatrix} e^t & 0 \\ 0 & e^{-t} \end{pmatrix} (t \in R) \qquad \gamma = \begin{pmatrix} -1 & 0 \\ 0 & -1 \end{pmatrix}$$

and let A be the group of all diagonal matrices in G. Then $A = A_0 \cup \gamma A_0$, where A_0 is the subgroup consisting of all h_t. Similarly, define

$$k_\theta = \begin{pmatrix} \cos \theta & \sin \theta \\ -\sin \theta & \cos \theta \end{pmatrix} \qquad (\theta \in R)$$

and let B be the subgroup consisting of all k_θ. Note that $\gamma = k_\pi$.

Let G' be the regular set of G as before. Then if

$$x = \begin{pmatrix} a & b \\ c & d \end{pmatrix}$$

is an element of G, it is easy to verify that $x \in G'$ if and only if $|\text{Tr } x| = |a + d| \neq 2$. Put $A' = A \cap G'$, $B' = B \cap G'$ and

$$G_A = \bigcup_{x \in G} xA'x^{-1} \qquad G_B = \bigcup_{x \in G} xB'x^{-1}$$

Then G' is the disjoint union of G_A and G_B.

Now for each integer $m \geqslant 1$, G has (apart from equivalence) exactly one irreducible (nonunitary) representation σ_m of degree m. Its character is given by

$$\text{Tr } \sigma_m(h_t) = \frac{e^{mt} - e^{-mt}}{e^t - e^{-t}} = (-1)^{m-1} \text{Tr } \sigma_m(\gamma h_t)$$

$$\text{Tr } \sigma_m(k_\theta) = \frac{e^{im\theta} - e^{-im\theta}}{e^{i\theta} - e^{-i\theta}}$$

On the other hand, the character $\Theta_m (m \neq 0)$ of the discrete may be described as follows.

$$\Theta_m(h_t) = \frac{e^{-|mt|}}{|e^t - e^{-t}|} = (-1)^{m-1} \Theta_m(\gamma h_t)$$

$$\Theta_m(k_\theta) = -(\text{sign } m) \frac{e^{im\theta}}{(e^{i\theta} - e^{-i\theta})}$$

There is a remarkable resemblance between the two formulas.

Similarly,

$$T_\lambda^+(h_t) = T_\lambda^-(h_t) = \frac{e^{i\lambda t} + e^{-i\lambda t}}{|e^t - e^{-t}|}$$

$$= T_\lambda^+(\gamma h_t) = -T_\lambda^-(\gamma h_t)$$

and

$$T_\lambda^+(k_\theta) = T_\lambda^-(k_\theta) = 0$$

for the characters of the continuous series. Finally,

$$T_0^- = \Theta_0^+ + \Theta_0^-$$

where $\Theta_0^\pm$ are irreducible characters given by

$$\Theta_0^\pm(h_t) = |e^t - e^{-t}|^{-1} = -\Theta_0^\pm(\gamma h_t)$$

$$\Theta_0^\pm(k_\theta) = \mp(e^{i\theta} - e^{-i\theta})^{-1}$$

Sequent Correlations in Evolutionary Stochastic Point Processes

S. K. SRINIVASAN[†]

INDIAN INSTITUTE OF TECHNOLOGY
Madras, India

1. INTRODUCTION

In this discussion, I propose to give a brief outline of some of the recent developments in what are called "stochastic point processes." I am afraid that the treatment may not be satisfactory to mathematicians, since our approach will be completely devoid of that mathematical tenor that is usually associated with stochastic processes. As Professor Ramakrishnan has observed, there is a group of mathematically oriented scientists who believe that the theory of stochastic processes is a part of measure theory and that, as such, it must be given a fair treatment. I wish to apologize for presenting an "unfair" treatment, that of resorting to phenomenological methods, with emphasis on techniques and mode of application rather than on abstract theory. Since the results are essentially new and the applications quite numerous, it is worth while to discuss the techniques in a purely phenomenological way.

With this preamble, let us consider a stochastic process progressing with respect to a continuous parameter t characterizing the process. Let us assume that the process involves another continuous parameter x with which a discrete random variable is associated; x may stand for the energy of a neutron if we are dealing with neutron transport theory, the age of a member of bacterial growth, or

---

†Department of Mathematics.

the energy of a cosmic-ray particle. The most natural way to deal with such a process is to think of the discrete variable $N(x, t)$ as a random variable whose properties should be connected to any information that we seek. However, in view of the continuous nature of the parameter x, we encounter a difficulty due to the situation that a comprehensive probability frequency function governs the distribution of $N(x, t)$. For this reason, such processes are known as *point processes* (see, for example, Bartlett[1]).

These processes are studied by defining certain density functions expressing correlations in x-space by Kendall[3] and Ramakrishnan[4] in entirely different contexts. Kendall, while dealing with the problem of age distribution in a population growth, introduced what he called "cumulant functions," which express the mean and covariance of the number of individuals whose age is distributed over certain infinitesimal intervals. Ramakrishnan, who dealt with the fluctuation in the number of electrons in an electron-photon cascade, introduced certain density functions now known as *product densities*. If $N(x, t)$ is the stochastic variable representing the number of entities at t with parametric value $X \leqslant x$, then we can consider the mean value of functions such as $dN(x, t)$, $dN(x_1, t)dN(x_2, t)$, etc. We shall presently discuss these things in detail.

The point of interest is that if we take t to be the time parameter, the correlation functions can be called *instant correlations,* since the correlations are defined for a particular t. I wish to emphasize the word *instant,* since in an entirely different context, Ramakishnan and Radha[5] introduced "sequent product densities." They were primarily motivated by a paper by Weisskopf in 1936 on the densiy correlations of electrons. Though the emphasis in Weisskopf's paper was to a great extent on the logarithmic spread of charge, the concept of density operators could be extended to different times. The application of such sequent densities leads to an alternative method of deducing the Feynman propagator for such systems.

To get back to sequent product densities, we notice that we can extend these ideas and define a family of such sequent correlations. To be very specific, let us consider a cascade process. If we take for x the energy of the particle, we can think of the energy x_1 at the point of production t_1† of the particle and the energy x_2 at a later

---

†The production can only be located in an interval $(t_1, t_1 + dt_1)$. so the correlation function will be a density function in t as well.

point t_2. (Here t stands for the depth.) Thus, we study the correlation between the "primitive" energy of the particle and its energy at any other depth. We note that this type of correlation leads to a density function defined in the product space $dx_1\, dt_1\, dx_2$. We shall discuss this point in more detail presently.

2. PRODUCT DENSITIES

Let $N(x, t)$ be the stochastic variable representing the number of entities at t with parametric values $X \leqslant x$. Then $dN(x, t)$ represents the stochastic variable representing the number of entities having parametric values in the elementary range dx. We shall assume that the probability that there occurs one entity in dx is proportional to dx, while the probability that there occurs more than one, say n, is of order $(dx)^n$. Thus it is possible to define a function $f_1(x, t)$ such that

$$f_1(x, t)dx = \mathscr{E}\{dN(x, t)\} \qquad (2.1)$$

where $\mathscr{E}$ denotes the average number of entities in dx. Defining $p(n)$ as the probability that there are n entities in dx, we have

$$\left.\begin{aligned}
p(1) &= f_1(x, t)dx + o(dx) \\
&= \mathscr{E}\{dN(x, t)\} + o(dx) \\
p(n) &= o(dx)^n \\
p(0) &= 1 - f_1(x, t)dx + o(dx)
\end{aligned}\right\} \qquad (2.2)$$

Thus the moments of $p(n)$ are given by

$$\mathscr{E}\{n^m\} = \sum n^m p(n) = \mathscr{E}\{[dN(x, t)]^m\}$$
$$= \mathscr{E}\{dN(x, t)\} = \mathscr{E}\{n\} \qquad (2.3)$$

Therefore, every one of the moments of the stochastic variable $dN(x, t)$ is equal to the probability that the stochastic variable takes the value 1.

With the help of these assumptions, it is easy to obtain a formula for the moments of the number of entities with parametric values lying in a given interval (x_l, x_u). Thus the mean number of entities is given by

$$\mathscr{E}\{N(x_u, t) - N(x_l, t)\} = \mathscr{E}\left\{\int_{x_l}^{x_u} dN(x, t)\right\}$$

$$= \int_{x_l}^{x_u} \mathscr{E}\{dN(x, t)\}$$

$$= \int_{x_l}^{x_u} f_1(x, t)dx \qquad (2.4)$$

while the mean square number is given by

$$\mathscr{E}\{[N(x_u, t) - N(x_l, t)]^2\} = \int_{x_l}^{x_u}\int_{x_l}^{x_u} \mathscr{E}\{dN(x_1, t)dN(x_2, t)\} \qquad (2.5)$$

In the integrand in (2.5), if dx_1 and dx_2 overlap, then by (2.3) it reduces to a single integral, the integrand being $\mathscr{E}\{dN(x, t)\}$ Thus

$$\mathscr{E}\{[N(x_u, t) - N(x_l, t)]^2\} = \int_{x_l}^{x_u}\int_{x_l}^{x_u} \mathscr{E}\{dN(x_1, t)dN(x_2, t\}$$

$$+ \int_{x_l}^{x_u} f_1(x, t)dx \qquad (2.6)$$

where $dx_1 \neq dx_2$. The integrand in first of the two integrals can be interpreted to be the simultaneous probability of finding two entities in the parametric values in the nonoverlapping ranges $(x_1, x_1 + dx_1)$ and $(x_2, x_2 + dx_2)$ at t. This is defined to be the second-order product density magnitude, the product density itself being given by

$$f_2(x_1, x_2, t)dx_1dx_2 = \mathscr{E}\{dN(x_1, t)dN(x_2, t)\} \qquad (2.7)$$

In a similar way product densities of higher order can be defined, and they will occur in expressions for higher moments of the number distribution.

From (2.6) it is clear that the mth moment will contain in the integrand product densities of degree less than or equal to m. An explicit expression can be obtained if we take into account the various ways in which degeneracies in the parametric values occur. For example, the contribution to a product density of degree s ($s < n$) will arise from various confluences of $n - s$ infinitesimal intervals, the maximum order of any particular confluence being $n - s$. The weight coefficients can be evaluated by combinatorial methods[†] and the mth moment can be written as

$$\mathscr{E}\{[N(x_u, t) - N(x_l, t)]^m\}$$

$$= \sum_{r=1}^{m} C_r^m \int_{x_l}^{x_u}\int_{x_l}^{x_u}\ldots\int_{x_l}^{x_u} f_r(x_1, x_2, \ldots, x_r, t)dx_1dx_2\ldots dx_r \qquad (2.8)$$

---

†An interesting way of arriving at these coefficients from a simple Poisson process is discussed in the paper of Ramakrishnan.[4]

where f_r represents the product density of degree r of entities. The moment formula is extremely useful in all physical situations. We can also extend this method to obtain correlations in the number corresponding to different t values and this is discussed by Srinivasan and Vasudevan.[12]

3. PRODUCTION PRODUCT DENSITIES

So far, we have not paid any attention to the nature of the parameter x. It may be called the *intrinsic parameter* for obvious reasons. If x denotes the age of an individual in a birth and death process, the age is a deterministic function in the sense that if the time t_0 of birth is known, the age of the individual can be specified at any later time $t > t_0$, the parameter t standing for time in this case. However, there are a number of physical processes where the time, or position in t-axis, of a creation of a particle and its x-value then do not determine its subsequent x-value.

A concrete example is provided by the energy state of an electron in the electron-photon cascade. In this case, the multiplicative stochastic process under consideration evolves with respect to both t and E, the energy parameter. In fact, it makes sense to talk of the parametric value at the point of the creation of the particle. Experimentally, it is convenient to make energy measurements at the point of production. As a matter of fact, the data on electromagnetic cascades observed in nuclear emulsions make reference only to the energies of the pairs of electrons at the point of production. Thus, we can talk of product density in the product space of t and E. Such product densities have been used in the formulation of cascade theories.

To define such a product density in the product space of x and t, we define $M(x, t)$ as the random variable representing the number of entities which are "born" at a parametric value $T \leqslant t$, the intrinsic parametric value being less than or equal to x. Then $dM(x, t)$ is the stochastic variable representing the number of entities that are created between t and $t + dt$ and have an intrinsic parametric value between x and $x + dx$ at its inception. If $\pi(n)$ is the probability that n entities are born between t and $t + dt$ with intrinsic parametric values between x and $x + dx$, then

$$\pi(1) = f_1(x, t)dxdt + o(dxdt) = \mathscr{E}\{dM(x, t)\}$$
$$\pi(0) = 1 - f_1(x, t)dxdt + o(dxdt)$$
$$\pi(n) = o(dxdt) \tag{3.1}$$

where $n > 1$.

From this point onward, it is clear that product density techniques for the product space Ω of x and t can be carried in toto. The moment formula (2.8) can be written for the product space by replacing dx by $dxdt$.

From the above, it is clear that once the product densities are known, the problem of fluctuations involves only integration over the parametric range. Thus, the main task in any problem will be the explicit determination of the product densities. The method of obtaining product densities for the case of electron-photon cascades has been dealt with by Ramakrishnan. However, a glance at the Mellin transform solution of the product densities will convince us of the gigantic magnitude of the task of anyone who attempts to obtain the same numbers from those results.

Calculation of higher moments becomes, though not impossible, formidably difficult. To offset this difficulty, there have been attempts to reformulate the problem in such a way that the evaluation of the cumulants (factorial moments) of the number distribution is somewhat less tedious (see for example Srinivasan[9]). This method completely eliminates the use of product densities or any other type of correlation functions and will not be discussed any further. Alternatively, we can introduce new density functions which include more information. Thus, in terms of those functions, it may not be necessary to go beyond the second order in any physical situation. For example, in the case of electromagnetic cascades, we can introduce two point correlations in E-space. Specifically, we can deal with particles that have been produced between t and $t + dt$ with energy lying between E and $E + dE$ at the point of production and are found to have an energy between E' and E' and $E' + dE'$ at $t' > t$.

Such a generalization, apart from the possibility of overcoming certain computational difficulties mentioned above, may be interesting from the point of view of comparison with experiments, particularly in cascades. Moreover, there are certain physical features characterizing the particles, an example being polarization of electrons or μ-mesons, which is directly related to the energy at the point of production rather than at a later point of observation. These new densities associated with more than one interval of parametric space may be called *evolutionary sequent correlations* for obvious reasons. We shall deal with these in the next two sections.

4. SEQUENT PRODUCT DENSITIES

The concept of sequent product densities was first introduced by Ramakrishnan and Radha,[5] who distinguished between "instant" and "sequent" correlations in point processes. The instant correlation relates to the study of correlations of the random variables corresponding to the same value of t, while the sequent correlation relates to that between the random variables corresponding to different values of t. Though the arguments used by Ramakrishnan and Radha depend heavily on t being the time parameter, the results are applicable to any ordered parameter. In fact, if $N(x, t)$ in the notation of section 2 is the stochastic variable representing the number of entities having a parametric value $X \leqslant x$, then the sequent product densities can be obtained by considering the expectation value of the product $dN(x_1, t_1)dN(x_2, t_2)\cdots dN(x_m, t_m)$. The sequent product densities contain more information than the instant product densities in that they partly explain the dependence of $N(x_i, t_i)$ on $N(x_j, t_j)$. Thus, the second-order sequent density is always expressible only in terms of second-order instant density.

We shall illustrate this by considering the nuclear cascade. Taking one type of particle only, the nth-order sequent density corresponding to different thickness is connected to all the instant densities of order m less than or equal to n. We first establish in the case of cascade processes the connection for the second order. For $t_2 > t_1$,

$$F_2(E_1, E_2; t_1, t_2 | E_0) = \int_{E_2}^{\infty} f_2(E_1, E_2' | E_0; t_1)F_1(E_2 | E_2'; t_2 - t_1)dE_2'$$

$$+ f_1(E_1 | E_0; t_1)F_1(E_2 | E_1; t_2 - t_1) \qquad (4.1)$$

where the parameter E_0 merely indicates that at $t = 0$ there was a primary energy of E_0. This equation is obtained by the following argument. The particle of energy E_2 at t_2 must belong to a cascade generated either by a particle of energy E_1 at t_1, or by the particle of energy $E_1(>E_2)$ at t_1. The two terms in (4.1) refer to these two possibilities. It is to be noted that F_2 is not symmetric in t_1 and t_2 as should be expected from the evolutionary nature of the Markovian process.

Defining $N(E, t_1)$ and $N(E, t_2)$ as the random variables representing the number of particles above energy E at t_1 and t_2 respectively, the expectation value of their product in given by

$$\mathcal{E}\{N(E, t_1)N(E, t_2)\} = \int_E^{E_0} \int_E^{E_0} F(E_1, E_2; t_1, t_2)dE_1 dE_2 \qquad (4.2)$$

The variation of $\mathcal{E}\{N(E, t_1)N(E, t_2)\}$ as t_1 varies from 0 to t_2 is not only of considerable mathematical interest but also of physical interest. For, if we assume the entire cascade be generated by a single particle of energy E_0, the second-order density at $t_1 = 0$ is zero and $f_1(E_1, 0) = \delta(E_0 - E_1)$. Hence, when t_1 approaches 0 the first term of (4.1) vanishes and F_2 reduces to

$$F_2(E_1, E_2; 0, t_2) = \delta(E_1 - E_0)f_1(E_2|E_1; t_2) \qquad (4.3)$$

As $t_1 \to t_2$, $F_2(E_1, E_2; t_1, t_2)$ does not reduce just to $f_2(E_1, E_2; t_2)$ but to

$$F_2(E_1, E_2; t_2, t_2) = \int f_2(E_1, E_2'; t_2)\delta(E_2 - E_2')dE_2' + f_1(E_1, t_2)\delta(E_2 - E_1)$$

$$= f_2(E_1, E_2; t_2) + f_1(E_1, t_2)\delta(E_2 - E_1) \qquad (4.4)$$

Hence, $\mathcal{E}\{N(E; 0)N(E; t_2)\}$ in the two limiting cases are given by

$$\mathcal{E}\{N(E; 0)N(E; t_2)\} = \int_E^{E_0} \int_E^{E_0} \delta(E_1 - E_0)f_1(E_2|E_1; t_2)dE_1 dE_2$$

$$= \int_E^{E_0} f_1(E_2|E_0; t_2)dE_2 \qquad (4.5)$$

and

$$\mathcal{E}\{[N(E, t_2)]^2\} = \int_E^{E_0} \int_E^{E_0} f_2(E_1, E_2; t_2)dE_1 dE_2 + \int_E^{E_0} f_1(E_1; t_2)\,dE_1 \qquad (4.6)$$

In the case of instant product densities, the rth moment is connected to all moments of order less than r in view of the degeneracies that occur in the density of degree r when the energy variables become equal. It turns out that this is implied in the definition of sequent product density and hence the limiting process $t_1 \to t_2$ in (4.2) gives the mean square number directly and involves the first moment also according to (4.6).

5. EVOLUTIONARY SEQUENT CORRELATIONS

We next proceed to a proper definition of evolutionary sequent correlations. Toward that end, let us define the primitive parametric value of an entity as the parametric value at its point of inception. We can consider the random variables $M(x_1, t_1; x_2, t_2)$ to represent the number of entities produced between 0 and t, the primitive parametric value of each of which is greater than or equal to x_1 the entities having a parametric value greater than or equal to x_2 at $t = t_2$. Then we can define an evolutionary sequent correlation density by considering $dM(x_1, t_1; x_2, t_2)$ which denotes the stochastic variable representing the number of entities that are produced between t_1 and $t_1 + dt_1$, with the primitive parametric value of each of which is between x_1 and $x_1 + dx_1$, the parametric value of these entities lying between x_2 and $x_2 + dx_2$ at $t = t_2$. We shall reserve the symbol $\mathcal{F}$ to denote such a product density. If $P(n)$ is the probability that the random variable $d\mathcal{M}$ takes the value n, it is reasonable to assume that $P(1)$ is of the order $\delta\Omega$ while $P(n)$ for $n \geqslant 2$ is of a smaller order of magnitude as compared to $\delta\Omega$, $\delta\Omega$ being an infinitesimal element in the space in which the density functions are defined.[†]

Thus, we can define the sequent correlation density of degree one by

$$\left.\begin{aligned}
P(1) &= \mathcal{F}_1(x_1, t_1; x_2, t_2)\delta\Omega + o(\delta\Omega) \\
&= \mathcal{E}\{d\mathcal{M}(x_1, t_1; x_2, t_2)\} \\
P(0) &= 1 - \mathcal{F}_1(x_1, t_1; x_2, t_2)\delta\Omega + o(\delta\Omega) \\
P(n) &= o(\delta\Omega)
\end{aligned}\right\} \tag{5.1}$$

where $n > 1$.

Higher order sequent correlation densities are defined in a similar manner. Correlation density of degree n is defined by

$$\mathcal{E}\{d\mathcal{M}(x_1, t_1; x_1', t)d\mathcal{M}(x_2, t_2; x_2', t)\ldots d\mathcal{M}(x_n, t_n; x_n', t)\}$$

$$= \mathcal{F}_n(x_1, t_1; x_2, t_2; \ldots; x_n, t_n; x_1', x_2', \ldots, x_n', t) \tag{5.2}$$

$$\delta\Omega_1\delta\Omega_2 \ldots \delta\Omega_n$$

provided the $\delta\Omega_i$ values are disjoint. If all the $\delta\Omega_i$ are not disjoint, then a degeneracy as in the case of usual product densities occurs. We shall not discuss this any further, since all the steps leading to the moment formula (2.8) are applicable in the present case provided dx is replaced by $d\Omega$.

[†] Ω is the product space of x_1, x_2, and t_1.

6. PHYSICAL APPLICATIONS

These types of product densities have been found useful in many physical phenomena involving some random structure. The first application has been in the field of electromagnetic cascades, and, in fact, general results on point processes have been obtained only from this angle. For a lucid account of the various methods that have been employed in cascades, the reader is referred to the work of Ramakrishnan.[7] Apart from this, product density techniques have also been utilized in the formulation of polarization cascades (see, for example, Ranganathan and Vasudevan[8]). In biology, we encounter a number of problems which can be formulated in a precise way by using the density function technique. Particular mention must be made of the theory of carcinogenesis, in which we need multi-point correlation functions. At the Rand Corporation, Bellman, Kalaba, and Vasudevan[2] have used the the product densities to deal with the energy-dependence of neutrons in transport phenomena.

As has been observed in the introductory remarks, the concept of sequent product density can be carried over into quantum mechanics. But in quantum mechanics, as contrasted with other situations we have described, there are the following features: (a) space–time symmetry and (b) condition on the surface of light cone, or, more specifically, causality conditions.

We shall take up condition (b). The role of causality in physical processes that are stochastic and nonstochastic in nature has been dealt with by Ramakrishnan.[6] The concept of causality is indeed difficult to express in terms of probability. To indicate the magnitude of the difficulty, let us consider the statement that B is caused by A. Can we assume that $P(B|A) = 1$?

Of course this means that B occurs if A occurs but does not imply $P(B|\text{non-A}) = 0$. In other words, besides A, there may be other events which cause B. The question is whether, given B, A is the event that has caused it. $P(A/B)$ is not determined by the information that $P(B/A) = 1$, a statement which is just equivalent to there being causes other than A for the occurence of B. Moreover, if we take into account the time parameter, this becomes more complicated because we have to deal with the concept of inverse probability, a notion which is not well understood at the moment.

Because of condition (a), it has been found advantageous to take the Fourier transform with respect to space-time and deal with the system in momentum space. In the latter case, the concept of time intervals loses all significance, and we try to connect only the state at the infinite past to the state at the infinite future. However, even here Watson and Goldberger have attempted to find a "coarse-grained" definition of "time intervals" from the S-matrix.† Moreover, the concept of sequent correlations is imbedded in the Green's functions, where they are written out in configuration space. In fact, these functions are utilized in the formulation of Caianiello to investigate the special features of quantum field theory. Moreover, the idea of confluence of two or more points in these correlation functions seems to play a very prominent role.

At this point, two general remarks should be made. The first is with respect to the functions $\mathscr{F}_1(x_1, t_1; x_2, t_2)$ defined in Ω, the product space of x_1, x_2, and t_1. On the other hand, it may be worth while to introduce the correlation between $\mathscr{M}(x, t)$, the random variable representing the number of entities that are created between 0 and t with primitive parametric values not less than x and the variable $N(x, t)$ representing the number of entities that are found at t with parametric values not less than x. Thus, we can deal with the function $F(x_1, t_1; x_2, t_2)$ defined by

$$F(x_1, t_1; x_2, t_2)\delta\Omega = \mathscr{E}\{d\mathscr{M}(x_1, t_1)dN(x_2, t_2)\}$$

Two interesting cases arise according as $t_1 > t_2$ or $t_2 > t_1$, and both the cases are of great importance in the interpretation of data on cosmic ray showers. Apart from this, $F(x_1, t_1; x_2, t_2)$ has some interesting limiting properties very similar to those of sequent product densities introduced by Ramakrishnan and Radha. The relevance of $F(x_1, t_1; x_2, t_2)$ and higher order correlations to evolutionary Markovian processes are of great interest from the point of view of stochastic point processes.

The other point is the manner of arriving at these various correlation functions in any particular physical situation. In fact, if the processes are Markovian, or at least in some sense quasi-Markovian in character, we can derive differential or integral equations for these correlation functions. I shall not go into these techniques,

---

†The author is thankful to Dr. Vasudevan for informing him of the results of Watson and Goldberger.

since they are far too involved to be discussed in the present context.†
However, there is one interesting feature that is characteristic of
cascade processes. These correlation functions are all, as they should
be, interrelated in the form of integral or differential equations.
However, we have recently observed that many of these correlation
functions could be shown to satisfy independent sets of equations
by the use of the invariant imbedding technique of Bellman.

7. SUMMARY

Stochastic point processes, which play a dominant role in many
physical phenomena, though very difficult to formulate in precise
mathematical terms, can be treated in a phenomenological way by
resorting to the correlation function techniques.

Once we start defining correlation functions, a number of multi-
point and sequent product density functions come to the fore.

These functions turn out to be precisely the quantities that are
of interest in physical phenomena such as cascade processes, shot
effect, Barkhausen noise, and possibly a variety of problems in biology.

It appears that these ideas may not be so fruitful in quantum
mechanics as they are in other branches. A full investigation of the
problem is necessary before a final verdict can be given.

However, so far as the general theory of stochastic processes is
concerned, these ideas have decisively proved to be advantageous,
though at the present moment they may not be entirely satisfactory
from the point of view of pure mathematics.

REFERENCES

1. M. S. Bartlett, *An Introduction to Stochastic Processes,* Cambridge University
   Press, 1955.
2. R. E. Bellman, R. Kalaba, and R. Vasudevan, *J. Math. Anal. Appl.* **8**: 225
   (1964).
3. D. G. Kendall, *J. Roy. Stat. Soc. B* **11**: 230 (1949).
4. A. Ramakrishnan, *Proc. Cambridge Phil. Soc.* **46**: 595 (1950).
5. A. Ramakrishnan and T. K. Radha, *Proc. Cambridge Phil. Soc.* **57**: 843
   (1961).

---

†For a derivation of the differential equations satisfied by these functions,
see Srinivasan and Iyer.[10]

6. A. Ramakrishnan, "Causality in Deterministic, Stochastic, and Quantum Mechanical Processes," preprint, 1962.
7. A. Ramakrishnan, *An Introduction to the Theory of Elementary Particles and Cosmic Rays*, Pergamon Press, London, 1962.
8. N. R. Ranganathan and R. Vasudevan, *Proc. Phys. Soc.* **76**: 650 (1960).
9. S. K. Srinivasan, in A. Ramakrishnan (editor), *Matscience Symposia on Theoretical Physics*, Vol. 2, Plenum Press, New York, 1966, p. 195.
10. S. K. Srinivasan and K. S. S. Iyer, *Nuovo Cimento* **33**: 273 (1964).
11. S. K. Srinivasan and K. S. S. Iyer, *Nuovo Cimento* **34**: 67 (1964).
12. S. K. Srinivasan and R. Vasudevan, *Nuovo Cimento* **41**: 101 (1966).

Author Index

A
Ademollo, 18, 19, 21
Amati, 91
Armenteros, 58
B
Balazs, 43
Bargmann, 98
Bartlett, 144, 154
Bell, 10
Bellman, 152, 154, 155
Beltrametti, 92
Bernstein, 10
Bethe, 120
Blankenbecler, 35
Blounin, 120
Born-von Karman, 121
Bouchiat, 18, 21
Bronzan, 58
Brown, 52, 57, 58
C
Cabibbo, 10, 21
Charap, 43
Chen, 120
D
Dalgarno, 120
Debye, 121
d'Espagnat, 21
Dietz, 59
Devanathan, 103
Dyson, 33
F
Feynman, 120
Frenkel, 127

Fubini, 19, 21, 43
Fujii, 104
Furlan, 19, 21
G
Gatto, 18, 19
Gelfand, 98
Gell-Mann, 21
Glashow, 85, 89
Goldberger, 104, 153
Gunson, 96
Gursey, 23, 25
Gyuk, 21
H
Hamilton, 57
Harish-Chandra, 98, 137
I
Iyer, 155
J
Jacobson, 122
K
Kacser, 51, 57
Kalaba, 152, 155
Kalbfleisch, 58
Kendall, 144, 155
Khuri, 58
Kuhn, 114
L
Landau, 120
Lander, 58
Lee, 10
Lifshitz, 120
Landsberg, 113
Lukierski, 69

Lurcat, 101
Luyton, 107
Luzzatto, 92

M
Meyer, 13
Mitra, 51, 58
Morgen, 113

N
Newton, 1, 2, 11
Nikitin, 58
Nishijima, 58
Nozières, 115, 120

P
Pais, 23
Prentki, 21
Pilkuhn, 63,
Pines, 115, 120
Primakoff, 104

R
Radha, 149, 153, 155
Radicati, 23
Ramakrishnan, 143, 144, 146, 148,
    149, 152, 153, 155
Ray, 58
Regge, 91
Reiche, 114
Rutherford, 2, 3

S
Sakita, 23, 25
Sakurai, 56, 58
Salpeter, 120
Samios, 58
Schiff, 58
Schlafli, 91

Scotti, 58
Schwartz, 137, 139
Schwinger, 32, 101
Sertorio, 97
Singer, 57, 58
Singh, 23
Sjolander, 121
Srinivasan, 143, 147, 148, 155
Stone, 131
Stueckelberg, 82, 89
Sugar, 35

T
Takahashi, 73, 89
Thomas, 114
Toller, 97
Treiman, 104
Tuan, 32

U
Umezawa, 73, 89

V
Van Hove, 125
Vasudevan, 147, 152, 155
Vankatesan, 91

W
Walker, 122
Wangler, 58
Watson, 153
Weinberg, 89, 104
Weisskopf, 1, 144
Wigner, 23, 25
Wong, 58

X
Xuong, 33

Subject Index

A

Acoustic phonons, 121
Application of the mass
    operators, 31
Artificial interaction of
    parameters, 119
Axioms and models, 131

B

Bethe–Salpeter equations, 40
Barkhausen noise, 154
"Bootstrap" theory, 8
Born-von Kármán equations, 126
Born-von Kármán theory, 126,
    127
Boltzmann constant, 127
Bragg scattering, 123
Brillouin zone, 124

C

Cascade process, 144
Casimir operators, 29, 30
Characters, 141
Circle group, 137
Coefficient of fractional
    parentage, 110
Coherent scattering, 123
Combinatorial methods, 146
Comments on sum rules, 113
Conditional projection, 74, 75
Conditional projection operators,
    74, 75
Conserved vector currents and
    broken symmetries, 13

Conserved vector current
    hypothesis, 9
Cosmic force, 10
Coupled channels, 37
Coupling constant, 16
Covariant causal propagator, 76
Covariant time-ordered product,
    76
CP invariance, 9, 10
Cumulant functions, 144

D

Debye–Waller factor, 126, 129
Density fluctuation operator, 115
Density matrix, 110
Dimensional analysis, 115
Dirac equations, 17, 105
Dirac measure, 137

E

Effective coupling constant, 15
Effective Hamiltonian, 104
Electromagnetic cascades, 148
Electron–photon cascade, 147
    148
Equivalent potential approach for
    strong interactions, 43
Evolutionary sequent correla-
    tions, 151
Extended Stueckelberg
    formalism, 82, 84

F

Feynman propagator, 144

Feynman's theorem, 114, 115
Fierz theory, 69
Fierz transformation, 9
Four-fermion interaction, 103
    104
Fourier series, 137

G
Gell-Mann–Okubo mass
    formula, 18, 26, 32
Generalization of Feynman's
    theorem, 117
G' parity, 19
Group representations for com-
    plex angular momentum, 91
Group SU_2, 14

H
Haar measure, 138, 140
Hypercharge symmetry, 5

I
Incoherent scattering, 123
Inelastic neuteron scattering and
    dynamics in solids and
    liquids, 121
Inelastic neutron scattering
    techniques, 122
Influence of the mass splitting,
    17
Instant correlations, 144
Instant and sequent correlations,
    149
Intrinsic parameter, 147
Invariant imbedding technique,
    154
Irreducible unitary, 141
Isospin symmetry, 3

J
$j-j$ coupling scheme, 111
Jordan–Pauli commutator, 73

K
Klein–Gordon equation, 75

L
Langevin's equation, 127, 129
Leptonic decay, 15, 18

Lie algebra, 13
$L-S$ or intermediate coupling
    scheme, 111

M
Many-body problems, 113
Markovian process, 149, 153
Mass sum rules, 26
Model of unitary S-matrix, 59
Mössbauer effect, 124
Multiplet assignment in $SU(6)$, 24
Multiplet structure, 23
Muon capture, 103, 104
Muon decay, 103, 104

N
Navier–Stokes equations, 122
N-electron atoms, 114
N-electron systems, 116
Neutron transport theory, 143
New approach to scattering
    theory, 35
Newton's *Opticks*, 1
Nozieres–Pines rule, 120

O
One particle exchange, 60
Optical phonons, 121

P
Pauli principle, 2
Periodic lattice, 114
Peripheral model, 60
Plancherel formula, 137, 141
Planck's constant, 115
Point process, 144
Product densities, 144, 145
Projection operators, 69
p-wave interaction, 53, 55

R
Raman scattering, 122
Rarita–Schwinger Lagrangian
    theory, 69, 74
Relativistic considerations, 40
Repulsive potential approach to
    pion resonances, 51
Results on liquid dynamics, 127

Results on phonons in solids, 126
Rotational symmetry, 2
Rydberg, 3

S

Schrödinger equation, 53, 54, 55
    116
Semi-simple lie group, 138
Sequent product densities, 144
Shot effect, 154
Single-channel scattering, 36
Stochastic point process, 143
Strangeness changing current, 13
Strangeness-conserving weak
    vector coupling constant, 16
Sum rules of the first kind, 117
SU_3 symmetry, 6, 8, 54, 55
SU_6, 6
Symmetry-breaking interaction,
    13
Symmetry-breaking Lagrangian,
    19
Symmetry under permutation, 2
s-wave interactions, 51

T

Theory of inelastic scattering of
    neutrons, 125
Thomas–Kuhn sum rule, 119
Topological group, 137
Types of spectroscopy, 5

U

Unconditional projection
    operator, 72

V

Van der Waals Force, 8
Vector coupling constant, 18, 19
Velocity dependent forces, 114
Velocity operator, 115
Virial theorem, 115, 116

W

Weak interactions, 5, 8, 9
Wigner–Eckart theorem, 107

Y

Yukawa interactions, 3